D1163814

SUPER VISION

SUPER VISION

A NEW VIEW OF NATURE

IVAN AMATO

FOREWORD BY
PHILIP MORRISON

HARRY N. ABRAMS, INC., PUBLISHERS

Project Manager: Eric Himmel
Editor: Sharon AvRutick
Designer: Helene Silverman
Photo Researcher: David Savage
Production Manager: Jane Searle

Chart on pages 11–14 designed by Agnieszka Gasparska

Library of Congress Cataloging-in-Publication Data

Amato, Ivan.
  Super vision : a new view of nature / Ivan Amato ; foreword by Philip
Morrison.
    p. cm.
Includes index.
  ISBN 0-8109-4545-2 (hardcover)
  1. Science—Pictorial works. 2. Science—Popular works. I. Title.

Q161.7.A43 2003
502.2'2—dc21

2003005117

Copyright © 2003 Ivan Amato

Published in 2003 by Harry N. Abrams, Incorporated, New York.
All rights reserved. No part of the contents of this book may
be reproduced without written permission of the publisher.

Printed and bound in Singapore

10 9 8 7 6 5 4 3 2 1

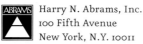 Harry N. Abrams, Inc.
100 Fifth Avenue
New York, N.Y. 10011
www.abramsbooks.com

Abrams is a subsidiary of

 LA MARTINIÈRE
G R O U P E

previous pages: See page 162.

page 5: See page 149.

pages 6–7: **City Lights** This image, a
composite of hundreds of images obtained
by satellites of the Defense Meteorological
Satellite Program (which is run by the U.S.
Department of Defense) in 1994 and 1995,
reveals what Earth looks like at night these
days. The oceans remain jet black, a reminder
that the majority of our planet's surface
remains untamed and unlit, and large areas
of land also remain largely dark at night.
The industrialized regions of the world stand
out, with Japan and the northeastern
United States shining upward with the most
intensity. By studying how the lighting
patterns change over time, researchers can
discern such large-scale phenomena as how
cities and regions grow and modernize.
The circumference of Earth, the entirety of
which is shown here, is about 40,000
kilometers (4×10^7 meters).

for Simon and Maxwell

CONTENTS

FOREWORD

Since the 1970s and 1980s, when my wife and I worked with the office of Charles and Ray Eames on both the film and book versions of *Powers of 10*, an examination of the sizes of objects in the universe and their relative scale, scientific imagery has only become more varied and revealing. What's more, it is increasingly blurring the lines between two Grand Categories, science and art.

As Ivan Amato proves in this striking book, the art is manifest: He presents his illustrations as beautiful, for so they are. Today's images, boosted with eye-catching color, stream past as you turn pages; they dazzle with a display of large and small structures or events.

Yet the images here are certainly not freed from science, from the meaning that they impart to the world. Indeed, one of Amato's strongest holds upon science is shown in his careful presentation of the actual size of every object imaged. When the image is so abstract that it refers to no particular scale of size, it enters the domain of modern science almost directly, for science itself images abstractions as forms of analysis. Thus many pictures of some scientific value resemble the work of abstract artists during many epochs of art.

Look at the image on page 38, a mandala, as Amato calls it with a poet's touch. Its strong symmetry, so common in artistic design, belongs to the micro-architecture of the unusual ceramic sample. This picture, like all the innumerable diffraction patterns recorded from electron scattering, displays not the ordinary space in which the atoms are so well arrayed, but a kind of space mathematically determined by the waves used in its making and the overall assemblage.

No part of the image represents any single atom. It is not a map of the crystal, but a display of its design as a whole, those orderly relationships in space much repeated in the huge number of unit cells of the sample, the basic motifs of nature's glorious penchant for order. This is a map of an abstract world of geometric relationships, a distinct richness held within the space in which we dwell, but hardly part of it. Its form and pattern, its size: all moot. It's art if you will, yet in no way free, save for its rather accidental size on the page and its pleasingly supplied colors.

In this volume, art is presented in long arrays of independent works, as in a museum. And as a book, each copy is similar to its fellows, the same design multiplied. The same is true for the arts of smaller cultures. Everyday ornament, say a fine pot or fabric, may be repeated in multiple, making a style of a single impetus. Science too deals in multiples: Stars and blossoms, the unending system of integers, the barrage of particle and photon, the sound trains of air vibration.

Amato's collection shares the property of multiplicity as well. Living forms are many, like the tiny viruses that add up to bring down a single organism, a hundred-pound human patient, or one bacterial cell, itself inexorably forced to multiply its infectious parasites and the billion cells that dwell in the same test-tube.

Super Vision is a treasury for the eye, brought by hand, tools, and mind. Our primate species descends from some sort of little, but big-eyed, insectivore in the old forests, adapted to catch bugs with clever fingers, enjoying and profiting from color too much to work only by night, when all is gray. This book can show you what became of our ancient beginnings.

PHILIP MORRISON
Institute Professor, Emeritus, and Professor of Physics, Emeritus, Massachusetts Institute of Technology

The Orders of Magnitude:
A SENSE OF THE WHOLE SHEBANG

The nucleus of an atom is as incomprehensibly small as the universe is unimaginably immense. Yet the images in this book represent phenomena whose sizes span the forty-two orders of magnitude that define this enormous range of spatial scales—from a millionth of a billionth of a meter (10^{-15} meter) up to a hundred trillion trillion meters (10^{26} meter).

One way to develop a sensibility for this spectrum of sizes is to think of an atomic nucleus as a dot. Roughly speaking, a grid of 10,000 dots is to a single dot what a nucleus is to an entire atom containing a nucleus and surrounding electrons. That grid spans four orders of magnitude, from about 10^{-15} meter to 10^{-11} meter (an typical atomic diameter is about 10^{-10} m). Now imagine that this entire first grid of 10,000 dots as a single dot of another grid of 10,000 dots. This second grid corresponds to the spatial ranges from 10^{-11} meter to 10^{-7} meter. A microscope view of each dot in this second grid would reveal its 10,000 constituent dots.

Now keep going. Imagine a third grid, each of whose 10,000 dots is equivalent to the entire second grid—each dot in the third grid is equivalent of 10,000 x 10,000 dots, or 100 million dots. It corresponds to a spatial range from 10^{-7} meter to a 10^{-3} meter, from viruses to sesame seeds. Continue with this sequence of grids until you reach the tenth grid. Here each dot corresponds to a sizable galaxy, and the full grid corresponds to the distance to the farthest cosmic objects observed. Spill over just one more order of magnitude, the one corresponding to 10^{26} meters, and you've entered the context in which the universe is known to reside.

On each page of the main part of this book, you will find a color bar that will help key each image to its place in the whole shebang.

$10 \, {}^{-5}_{\mathrm{m}}$ $\quad$ $10 \, {}^{-4}_{\mathrm{m}}$

$10 \, {}^{2}_{\mathrm{m}}$ $\quad$ $10 \, {}^{3}_{\mathrm{m}}$ $\quad$ $10 \, {}^{4}_{\mathrm{m}}$

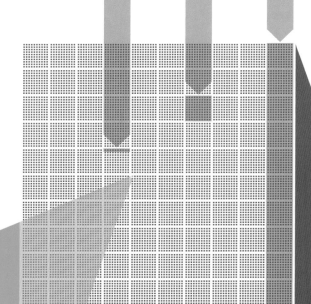

ENTIRE GRID: $10 \, {}^{1}_{\mathrm{m}}$

Length of a bus

$10 \, {}^{-3}_{\mathrm{m}}$	1 millimeter, the size of a pin's head
$10 \, {}^{-2}_{\mathrm{m}}$	1 centimeter, the size of a fingernail
$10 \, {}^{-1}_{\mathrm{m}}$	Width of a palm
$10 \, {}^{0}_{\mathrm{m}}$	1 meter, the height of a child

$0 \, {}^{-3}_{\mathrm{m}}$

rts such	
a cell's nucleus cells	
logical cells nd	
salt	

ENTIRE GRID: $10 \, {}^{5}_{\mathrm{m}}$

Distance traveled in one hour on a highway

$10 \, {}^{1}_{\mathrm{m}}$	Length of a bus
$10 \, {}^{2}_{\mathrm{m}}$	Length of a football field
$10 \, {}^{3}_{\mathrm{m}}$	1 kilometer, the size of neighborhoods and small villages
$10 \, {}^{4}_{\mathrm{m}}$	Distance from the top of Mt. Everest to the upper atmosphere

$10 \, {}^{-2}_{\mathrm{m}}$ $\quad$ $10 \, {}^{-1}_{\mathrm{m}}$ $\quad$ $10 \, {}^{0}_{\mathrm{m}}$

$10 \, {}^{-3}_{\mathrm{m}}$

$10 \, {}^{-3}_{\mathrm{m}} - 10 \, {}^{1}_{\mathrm{m}}$

$10 \, {}^{1}_{\mathrm{m}} - 10 \, {}^{5}_{\mathrm{m}}$

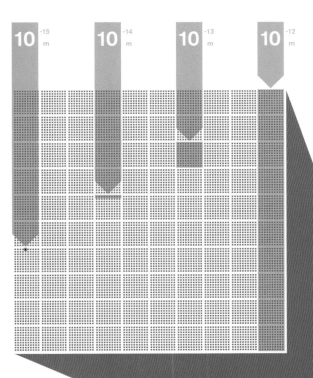

10⁻¹⁵ m   10⁻¹⁴ m   10⁻¹³ m   10⁻¹² m   10⁻⁶ m

ENTIRE GRID: 10^{-7} m

Cellular anatomical parts such as chromosomes

10^{-11} m	About 1/10 the diameter of a hydrogen atom
10^{-10} m	1 angstrom, the scale of individual atoms or small molecules
10^{-9} m	1 nanometer, the scale of larger molecules such as proteins
10^{-8} m	Diameter of some viruses and the thickness of some cell walls

ENTIRE GRID: 10^{-11} m

About 1/10 the diameter of a hydrogen atom

10^{-15} m	Diameter of a proton, the nucleus of a hydrogen atom
10^{-14} m	Diameter of a gold atom's nucleus (79 protons and 118 neutrons)
10^{-13} m	Length of 10 shoulder-to-shoulder atomic nuclei
10^{-12} m	1 picometer, the wavelength of gamma radiation

ENTIRE GRID:

1 millimeter, the size of a pin's head

10^{-7} m	Cellular anato... as chromosom...
10^{-6} m	1 micron, the s... and of small b...
10^{-5} m	Diameter of ty... and small grai...
10^{-4} m	Diameter of a ...

10⁻¹⁰ m   10⁻⁹ m   10⁻⁸ m

10^{-15} m — 10^{-11} m

10^{-11} m — 10^{-7} m

10^{-7} m —

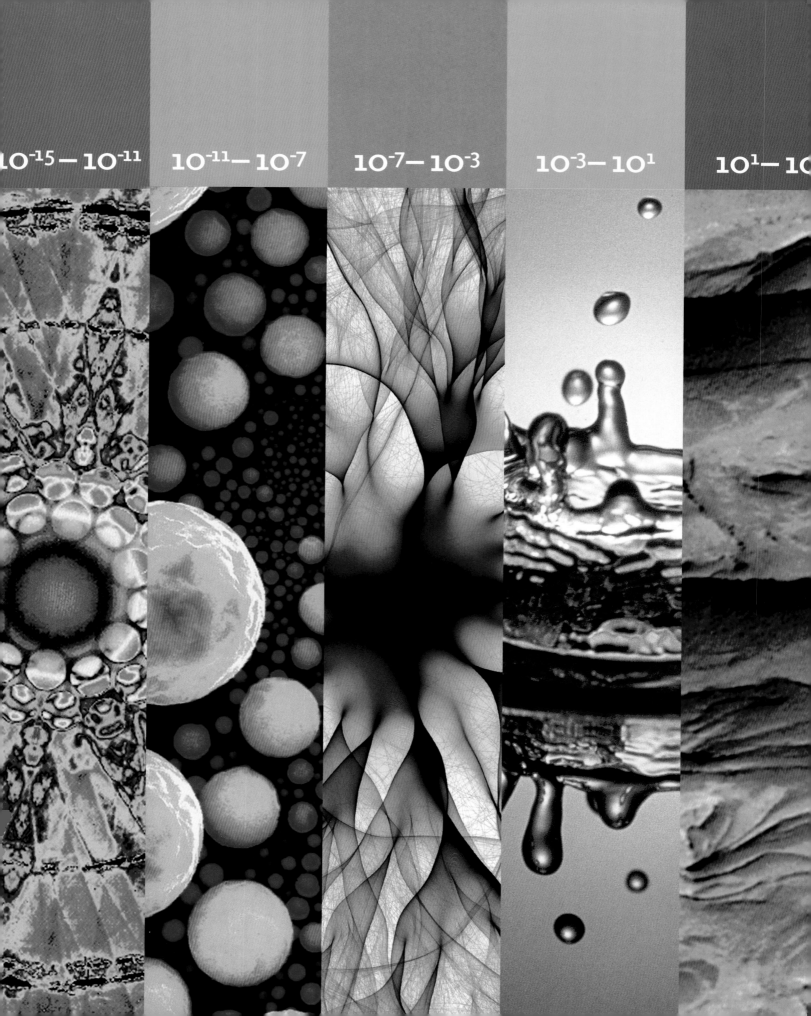

$10^{-15} - 10^{-11}$ $10^{-11} - 10^{-7}$ $10^{-7} - 10^{-3}$ $10^{-3} - 10^{1}$ $10^{1} - 10$

$0^5 - 10^9$ $10^9 - 10^{13}$ $10^{13} - 10^{17}$ $10^{17} - 10^{21}$ $10^{21} - 10^{25}$

10¹⁸ m 10¹⁹ m 10²⁰ m 10²⁶ m

ENTIRE GRID: 10^{25} m

Distance to quasars

10^{21} m	Diameter of the Milky Way Galaxy
10^{22} m	Distance to the Andromeda Galaxy
10^{23} m	Size of galactic clusters, such as our own Local Group
10^{24} m	Distance light travels in 100 million years

ENTIRE GRID: 10^{21} m

Diameter of the Milky Way Galaxy

10^{17} m	Distance to the star Sirius
10^{18} m	Thickness of the Milky Way
10^{19} m	Distance to the Horsehead Nebula
10^{20} m	Distance from Earth to the center of the Milky Way

10^{26} m

The largest structural scales of the known universe; about 13.7 billion light-years

10^{22} m 10^{23} m 10^{24} m

10^{17} m — 10^{21} m 10^{21} m — 10^{25} m 10^{26} m

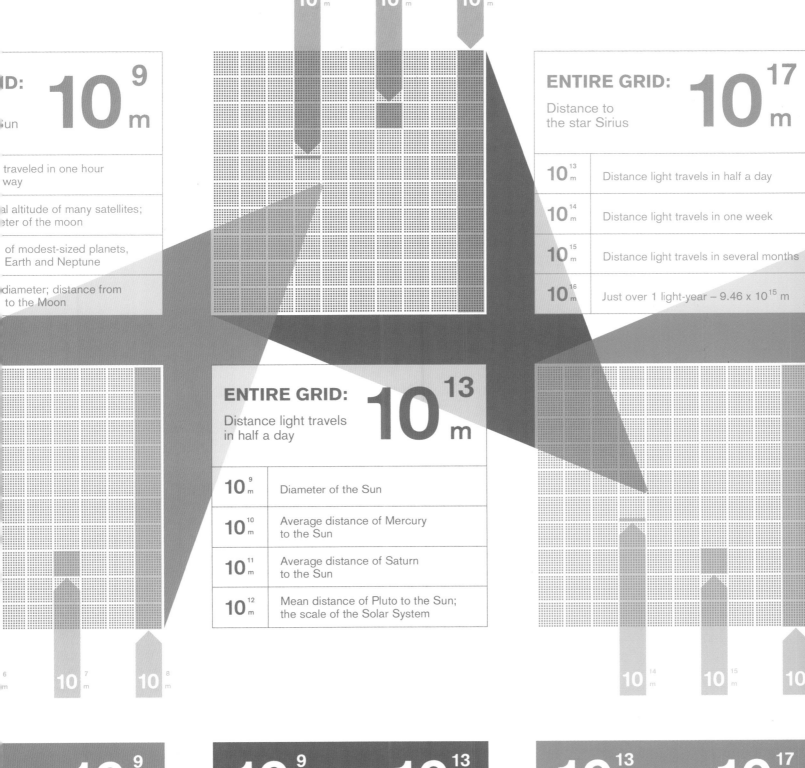

ID: 10^9_{m}

...Sun

...traveled in one hour
...way

...al altitude of many satellites;
...eter of the moon

...of modest-sized planets,
...Earth and Neptune

...diameter; distance from
...to the Moon

10^{10}_{m} 10^{11}_{m} 10^{12}_{m}

ENTIRE GRID: 10^{17}_{m}

Distance to
the star Sirius

10^{13}_{m}	Distance light travels in half a day
10^{14}_{m}	Distance light travels in one week
10^{15}_{m}	Distance light travels in several months
10^{16}_{m}	Just over 1 light-year – 9.46×10^{15} m

ENTIRE GRID: 10^{13}_{m}

Distance light travels
in half a day

10^9_{m}	Diameter of the Sun
10^{10}_{m}	Average distance of Mercury to the Sun
10^{11}_{m}	Average distance of Saturn to the Sun
10^{12}_{m}	Mean distance of Pluto to the Sun; the scale of the Solar System

10^6_{m} 10^7_{m} 10^8_{m}

10^{14}_{m} 10^{15}_{m} 10

— 10^9_{m} 10^9_{m} — 10^{13}_{m} 10^{13}_{m} — 10^{17}_{m}

PREFACE

For all but the most recent pages of the human story, to see nature has been to see the world as one's own eyes—and ears and nose and other sensory anatomy—could take it in. Although ancient rational minds proved they could reason their way to phenomena beyond human sensitivity, the perceivable reality of the world always has been circumscribed by the biological kit of sensory gadgetry that human beings bring with them through the birth canal.

The compulsion to record this sensory experience of nature seems to be as fundamental a part of being human as is using that sensory input to perceive what matters in the environment —attractive mates, sharp-clawed predators, the color and smell of nutritious fruit, the bitter taste of poison. Those famous Paleolithic cave paintings, such as in the Chauvet-Pont-d'Arc cave in France, where lions, bears, and sparring rhinoceroses grace the walls, provide hard evidence that the drive to depict nature was being expressed 30,000 years ago and earlier. That drive has never stopped.

What has changed utterly in the meanwhile, particularly in the past half-millennium, and most particularly in the last few decades, is that we now have the ability to assist and boost our raw senses with miraculously capable technologies. Just as those in Albert Einstein's and Max Planck's generations rewrote Newton's long-standing laws of physics as they analyzed those physical events that unfold at diminutive scales that were experimentally inaccessible to earlier scientists, today's generation of tool-wielding researchers are re-envisioning what nature looks like at scales and in modes that were once unimaginable. These new tourists of nature have given us the gift of Super Vision in a thousand different ways.

With radio telescopes, X-ray telescopes, light microscopes, electron microscopes, scanning probe microscopes, thermal detectors, X-ray crystallography machines, magnetic resonance imagers, 3-D sonography systems, ion probes, hyperspectral overhead imagers, gamma-ray spectroscopes, gravity detectors, fluorescence probes, infrared spectrometers, neutron activation analysis, and countless other instruments and analytical methods, we have flayed nature down to its subatomic rudiments and simultaneously revealed its unfathomable grandeur. With a scanning tunneling microscope, we can examine the seemingly endless atomic landscape of a fleck of nickel as though we were flying over it in a superlatively tiny helicopter. With the orbiting Hubble Space Telescope we routinely witness the fantastically rococo forms of stars, galaxies, and galactic clusters, once visible at best merely as points of light. If nature as perceived by our unassisted senses already is awesome, and it is, then our growing Super Vision of nature is supremely so.

Like the Paleolithic cave painters, the inventors and scientists who have been devising and using the tools behind this new way of envisioning the world are creating portraiture of nature. Go to their offices and labs, attend their conferences, read their journals, and you can't help but notice how image-driven so much of science has become. For most of the scientists themselves, these images serve primarily as information—graphic ways of testing theories, recording measurements, and gathering data.

Yet many of these same investigators evidently see something more. Why else would they frame their data and hang it on their walls? For what other reason would magazines like *Nature* and *Science* proudly display these images on their covers, in full color and from page edge to page edge?

This book is a celebration of our new, hard-won Super Vision of nature. The two hundred or so images on these pages are arranged roughly in size order, beginning with the fundamental particles that make up all matter and ending with galaxies and even much larger neighborhoods of the universe. Collectively, the images and their captions harbor an epic story of science, discovery, and the human compulsion to probe nature. These images are merely a tiny portion of those that the scientific community has produced, which means that this collection derives from an exercise of selection dominated by exclusion. The complete gallery of science imagery is truly gargantuan and getting larger by the second.

Each image that ultimately made it into this book moved me first on the aesthetic level. Simply put, I searched for pictures whose colors, forms, and composition drew my eye and kept it there. Some colors were not present in the raw images recorded from, say, an electron microscope. Instead, using image-processing tools, they were added digitally by the scientists themselves or by others, with the goal of highlighting specific features of the images or making them more visually striking. I also strove to assemble a portfolio that illustrates the astounding powers of observation wielded by today's armamentarium of scientific instruments. And I aimed to assemble a collection that would provide a pastiche of imagery reflective of the fantastically varied elements that mark the scientific way of telling stories about the world.

Quite often, after learning about the meaning and context of a specific image, it would move me in a second way, one that combines an intellectual and an emotional component. To see a gorgeous image of a red sphere rising on what appears to be the horizon of a strange ocean, and then to learn that it depicts a human immunodeficiency virus (HIV) emerging from an immune system cell as seen with an electron microscope requires me somehow to reconcile the pure beauty of nature with its ability to wreck life. To see our planet from the perspective of a spacecraft circling the moon is to be reminded that other intelligences may be looking Earthward just as we desperately look outward for signs of other sentient creatures.

The pictures also speak of new technologies that will transform how we live. To view a stunning dimpled vista that looks like an infinite egg carton, and then to realize it is a portrait of the arrangement of atoms in a piece of metal analyzed with a scanning tunneling microscope, is to look at what used to be hopelessly inaccessible. And using that same tool, known in short as an STM, we even can move individual atoms, almost as though we were using our own hands, and build molecules and other diminutive structures with them. If individual atoms have become our playthings, where can the line of technological impossibility be drawn?

Every picture in this book is an invitation to a startling realization of that kind, to an enriching reflection, and to a fresh new look at the ever-more-visible universe.

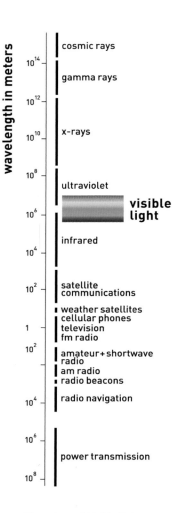

wavelength in meters

10¹⁴ — cosmic rays

gamma rays

10¹² —

10¹⁰ — x-rays

10⁸ — ultraviolet

10⁶ — **visible light**

infrared

10⁴ —

10² — satellite communications

1 — weather satellites
cellular phones
television
fm radio

10² — amateur+shortwave radio
am radio
radio beacons

10⁴ — radio navigation

10⁶ —

power transmission

10⁸ —

The colors of visible light comprise a mere sliver of the full electromagnetic spectrum.

One of the most fantastic truths that a sentient being can contemplate is that the universe, at some point over its multibillion-year evolution, cleaved into a part that can see and a part that can be seen.

And not just seen. The universe also became smellable, tastable, audible, touchable, and otherwise discernible by a living kingdom rife with sensory anatomy. The human manner of seeing the world is but one way of perceiving what's out there. There are many, many others.

The placement of a duck's eyes on its comedically narrow head gives it wraparound vision, enabling it to see a 360-degree vista of its environment. We need to twist our necks or turn around to see behind us.

Some vipers, including rattlesnakes, have a little pit between each eye and nostril. About as wide as a shish-kabob skewer and less than a quarter-inch deep, these pits contain a membrane riddled with thousands of heat-sensitive cells. With them, these creatures see infrared emanations from, say, a soon-to-be-eaten rodent—heat vision.

Some insects, including bees and butterflies, see in ultraviolet wavelengths. Perhaps not coincidentally, flowers often harbor ultraviolet pigments in arrangements that seem to say, "Land here!"

Bats rely on high-frequency sound to "see" in pitch blackness and, for all practical purposes, bloodhounds can envision landscapes in topographies painted in aroma—call these sonar vision and smellovision.

Some fish, including sharks and rays, can discern the electrical features of their surroundings. Specialized organs arranged on their heads pick up tiny electric fields generated by the moving muscles of nearby fish even in murky and muddy waters that otherwise render those fish invisible.

Our own sensory equipment serves us well enough, but it actually leaves us blind to most of the phenomena in the universe. The vast majority of what is out there is just too small, or too big, or the wrong color or sonic frequency, or the wrong type of energy, or too fast, or too slow, or otherwise beyond our ability to notice. It is as though we are looking out on the world from inside a box perforated by just a few peepholes, as the science writer Jillyn Smith once aptly put it.

Consider vision. The colors of the rainbow—our own visible spectrum—had always seemed unique, but in 1864, when James Clerk Maxwell first revealed his equations that describe electromagnetism, the colors of the rainbow were shown to amount to a tiny sliver of the vast electromagnetic spectrum. Go in one direction from the visible portion of the electromagnetic spectrum, and the last bit of violet fades out, giving way to ultraviolet colors, then to X-ray colors, then to ever more exotic invisible colors known as gamma rays. Go the other way and the last bit of red gives way to infrared colors, which we feel as heat instead of see as beyond-red colors. Continue in that direction and you get to the longer wavelengths that now fill the airwaves conveying radio and television programs, billions of cell-phone conversations and pager signals, and radar signals from air-traffic control towers and air-defense systems.

We are blind to far more than just most of the electromagnetic spectrum. We also are limited in what we can perceive even within the colors we can see. That's because the spatial scale of all things is so vast—spanning more than forty orders of magnitude from subatomic particles to the entire universe—that we cannot comprehend most of it. We either are giants unable to directly see the constituents of the things we hold in our own hands or mites who might think the entire world fits on a leaf.

The smallest speck of dust that we can see without a magnifying glass or microscope is about 10 microns (ten millionths of a meter—10^{-6} m), or about 1/250 of an inch. A typical human hair, which is visible enough, might have a diameter of about 75 microns. But we can't see the flaky structure of the cells that make up the hair, the molecules that make up these cells, or the atoms that make up those molecules.

lines 1 mm apart

Lines as closely spaced as one-tenth of one millimeter (100 microns) would be hard to discern without a good magnify-ing glass.

A single atom of hydrogen, after all, spans less than 1 angstrom, which is one ten-billionth of a meter (10^{-10} m). It's as much smaller than that minimally visible speck of dust as a poppy seed is shorter than a 100-meter-tall sequoia. And atoms themselves are made of parts that are even smaller, and for that reason, even more thoroughly invisible to our naked eyes. An atom's naked nucleus (the atom minus its electrons) is to the entire atom what that poppy seed is to the sequoia. The single proton that makes up most hydrogen nuclei spans about one-hundredth of a femtometer, which is to say it is one millionth of one billionth of a meter (10^{-15} m) in diameter. That's about 100,000 times smaller than the diameter of the complete atom, which includes an electron in addition to its nucleus.

We are not nearly as limited in our ability to visualize really big things, but we are not able to see them in much detail. The only way to visualize a skyscraper in its entirety, for example, is to stand far back from it. But when we reach a distance that enables us to see its outlines and its grosser architectural features, we lose the ability to see the finer texture of its facade. Conversely, if we get close enough to these details to see them, we lose sight of the entire building. It's the same as losing the forest for the trees, and it only gets worse as the things we look at get bigger.

To visualize the whole Grand Canyon, for example, we have to peer down from a plane or use satellite images. We can see the entire moon and the Sun because, in effect, we are flying over them at enormous altitudes. For the same reason, however, we see them only schematically, with almost no details. Anything bigger than the Sun is way beyond our solar system, which means it appears to our eyes only as little structureless points of light at best. Yet our universe contains roughly a hundred billion galaxies, each of which is formed of roughly a hundred billion stars. It is almost paralyzing to imagine just how exquisitely constrained is our view of the universe's portfolio of phenomena.

Our sensory box is pierced with other kinds of peepholes that let us hear and smell and taste and feel the world, albeit all with their own limitations. Ears are wondrous instruments for transforming vibrations in the air into neural signals that brains can interpret as sounds with particular meanings or as noise with no readily discernible significance. But like our eyes, our ears are sensitive to just a portion of the possible range of sound waves.

The adult human ear responds to vibrations as slow as about 20 per second and up to about 20,000 vibrations per second (20–20,000 hertz). The human voice, ranging from a deep bass to a shrill soprano, corresponds to frequencies of about 80–1000 hertz. Again, other members of the living kingdom have evolved abilities to sense sound beyond the human range of sensibility. Bats use high-frequency sound like a flashlight, sending out ultrasonic chirps up to 200,000 hertz and then listening for echoes from the various objects in their environment. When sound bounces off an object moving toward the bat, the echo returns with a pitch that is slightly higher (if the source is moving toward the listener); when that echo comes from something moving away, the returned echo is of a lower frequency. Using that kind of information, known as a Doppler shift, the bat can home in on prey on the blackest of nights. Some bats even catch fish by detecting the wakes they leave in the surface of a river or lake. These sensory capabilities have made bats—and their marine brethren who use natural sonar to navigate and hunt—objects of envy of military researchers who aim to develop technologies that can duplicate or surpass these capabilities.

People are completely insensitive to the ultrasonic chatter that goes on around them. Besides bats, many other animals and insects squeal and squeak, or their wings and other body parts vibrate, at ultrasonic frequencies inaudible to us. The airwaves are filled with a cacophony of inaudible sounds as rich and varied as the sonic montage of a city. But even that combination of vehicular sounds, voices, music, machines, and sirens contains another realm of sounds we cannot hear. Every jangling key has ultrasonic pitches that we are deaf to, and the same goes for screeching brakes, police officers' whistles, and silverware scraping across plates.

Then there are infrasonic waves whose frequencies are too low for us to hear. Pigeons can detect air vibrations as slow as one every ten seconds. That's .1 hertz. Other animals, including elephants and some birds, also hear infrasound. Elephants rely on this sensory peephole to communicate with each other over great distances. Some birds and other animals may use ultrasound to hear the infrasonic frequencies generated by thunderstorms too distant for a human being in the same place to hear.

Our peepholes on chemistry, mediated by our nose and our tongue, are impressive. We can detect tens of thousands of nutritious, poisonous, noxious, and alluring chemicals. These peepholes provide us with great pleasures and they help us avoid spoiled foods, fires, and other threats. But our chemical peepholes are rank novices compared to those of many other animals. Consider that the olfactory membrane in a person's nasal passages spans the size of a postage stamp, whereas a cat's olfactory membrane covers almost four times that area, and a bloodhound's can cover almost forty times as much. By virtue of its olfactory capabilities, a bloodhound is more than one million times more sensitive to the smell of a human being than a human being is. What's more, it can distinguish the subtly different cocktail of aroma molecules that emanate like a chemical fingerprint from each person. If we had such a nose, we could enter a silent but crowded hall and know where different people stood simply by sniffing. We even would know who had left the hall a half-hour ago.

Tactile sensations—among them pain, texture, temperature, moisture, and very low mechanical frequencies such as the rumbling of a truck going by—provide us for the most part with very localized information about the world, most often things we are holding in our hands or are otherwise touching. The same is true for other creatures, but their sensory capabilities often outperform ours. Scorpions have sensory hair cells on their pincers that respond to air currents a mere one-hundredth as powerful as the lightest breeze we can detect. Cockroaches have vibration sensors in their legs capable of detecting movements far less than the diameter of a bacterium. Many fish have so-called lateral line organs—sensory pits embedded along their head and bodies—with which they can detect differences in flow due to their surroundings, presumably painting a picture with brushstrokes made of small water currents.

Getting a feel for the limitations of our sensory peepholes only begins to characterize our blindness. There is another multifaceted kind of invisibility due to our absence of appropriate sensory anatomy. Magnetic fields, for one, remain invisible to most of God's creatures. An exception is the class of microbes known as magnetotactic bacteria. Within their tiny single-celled bodies, they manufacture even tinier crystals of magnetite and other magnetic minerals and arrange them, like tiny pearls, into diminutive chains. With these, the microbes, even after being churned up by currents, can sense the Earth's magnetic field and use that sensibility to navigate back to nutrient-rich sediments (they're too small for gravity to pull them down).

We have no way of directly seeing electricity either, or atmospheric pressure, or the crystal structure of metals, or the electrochemical flows in the brain, or what's behind skin or walls, or an infinitude of other things which we could conceive of seeing if only we had the means. The ways of blindness are many.

What's more, so much of what happens in this universe is unseeable by virtue of the speed at which it occurs. Even if we could see a molecule of water, consider that in a second, that molecule, which is shaped something like a V with an oxygen atom at the vertex and a hydrogen atom at the end of each leg, executes a frenetic hyperspeed dance. In the time it takes you to say "one-one-thousand," the molecule's two legs flex back and forth toward one another literally many trillions of times. Meanwhile, with each flex, the hydrogen atoms at the end of each leg twirl around the oxygen hub, like sub-nanoscopic dervishes. This is the temporal realm in which chemistry gets done and we have no natural peephole to experience it directly.

Some aquatic bacteria manufacture minuscule magnetic iron-based particles called magnetosomes that provide the microbes with sensitivity to Earth's magnetic field.

This hyperspeed time domain is also the realm in which matter emits light, absorbs it, and otherwise interacts with it. When a photon enters a light-sensitive rod cell on your retina and initiates a nerve-triggering mechanism by hitting a pigment in the cell known as rhodopsin, it happens invisibly in a billionth of the time it takes to blink an eye. Vision—the collective message from billions of photons triggering massive cascades of nervous impulses inward to a brain that interprets it all as a friend or a foe, as a danger or a delight—is the biological point. That we don't have a direct peephole for "seeing" the physics and chemistry underlying the process of seeing doesn't really matter when it comes to surviving and evolving. But it does exemplify the many natural and invented processes to which we are utterly blind.

So much to see. So few peepholes. That we are so supremely lucky to be among the part of the universe that can see itself is enough of a gift to stave off any real sense of deprivation. After all, on this entire planet, is there any other entity—animate or inanimate—that enjoys the privilege of self-awareness and the degree of knowledge about the world that we have? And, fantastically, we humans have discovered ways of creating new peepholes for ourselves. Over the past four centuries or so, the scientifically and technically minded among us have been inventing all manner of instruments that have been expanding our sensory access to previously invisible and even unimagined phenomena—from subatomic particles to viruses to gigantic stars in their very death throes.

THE IDEA of assisting the senses, of expanding our peepholes on the world in order to discern more than is directly apparent, undoubtedly happened in many places at many times. But it built up the unstoppable momentum that continues today in the early phases of the scientific revolution. This drive to enhance sensory access to the world has enabled scientists to explore known phenomena at dimensions beyond what their own senses could discern while opening novel windows onto phenomena previously unseen or even imagined. "The leap from naked-eye observation to instrument-aided vision would be one of the great advances in the history of the planet," wrote historian and former Librarian of Congress Daniel J. Boorstin.

This leap began in earnest with lenses in the hands of the likes of Galileo Galilei, who in the second half of 1609 first turned a newfangled military tool, the telescope, toward the heavens, and Antonie van Leeuwenhoek, a plain-speaking Dutch haberdasher who later in the century used a single-lens microscope to discover "animalcules," thereby founding the science of microbiology. With their common basis in lenses and optics, microscopes and telescopes took the scientific community in trajectories toward never-before-seen microscapes and macroscapes. Each trajectory has taken generation after generation of observers into parallel universes of wildly diverse phenomena— from bacteria to black holes. As Victor Hugo put it in *Les Misérables*, "Where the telescope ends, the microscope begins. Which of the two has the grander view?"

Every invention of a new instrument for enhancing human sensibility has been a portal to another world of phenomena. This was quite literally true when Galileo made the stunning discovery that Jupiter had moons of its own. He described it as another

Robert Hooke's drawing of a flea (Micrographia, 1665).

solar system, a powerful experimental sign that the ancient portrait of the universe with Earth at the center of a comprehensibly finite heaven—a picture mandated in the West by both religion and tradition—was seriously flawed. Early microscopists were so enthralled by the otherwordly perspective their instruments provided that some of them eagerly viewed whatever they could place under their lenses—razor blades, seeds, insects, as well as semen, pond scum, and feces. A flea became a monster with natural armor and intimidating built-in weaponry. The English natural philosopher Robert Hooke, an early microscopist as well as an accomplished draftsman, published many of his own drawings in 1665 in his *Micrographia*. So compelling were these images, including one of a flea, that readers of this work often cut the plates out and framed them, an early sign that scientific imagery's content—its role as data—could be presented in aesthetically pleasing ways.

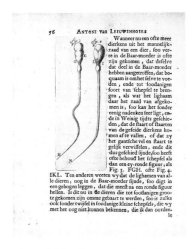

Through his microscope, Antonie van Leeuwenhoek spied sperm cells (above) and "multifarious animalcules."

Although his drawings were not nearly as beautiful as Hooke's, the Dutch microscopist Antonie van Leeuwenhoek, a friend of the artist Jan Vermeer, perfected the art of grinding single, ladybug-size lenses that for a while in the latter part of the seventeenth century enabled him to view the world on finer scales than had any other human being. In 1674, for example, he looked at a drop of water from a lake near his own home in Delft and saw teeming masses of "multifarious animalcules," which he described as the most marvelous discovery he had made yet. Wrote Leeuwenhoek in a scientific letter to the Royal Society in London, "I must say, for my part, that no more pleasant sight has ever yet come before my eyes than these many thousands of living creatures, seen all alive in a little drop of water." In 1673, Leeuwenhoek became the first biologist to describe spermatozoa.

Leeuwenhoek's microscopes would not be outperformed until the nineteenth century, when lens designers and instrument makers overcame some of the major optical flaws that had been ineluctably present in earlier instruments—such as the blurring produced when light enters highly curved lenses at different points and then bends to different degrees (spherical aberration), and the way the different colors comprising white light refract through lenses at different angles (chromatic aberration). Others would push microscopy forward not by improving lens systems, but by making specimens show their internal anatomy more readily. The Englishman Henry Sorby did this in the 1860s when he developed the process of metallography. He found that an acidic etching treatment of a sliced and polished metal specimen enabled a microscopist to view the size and relationships of the metal's constituent crystalline grains and noncrystalline inclusions. The Spanish anatomist Santiago Ramón y Cajal rendered visible the intricate forms of brain cells, as well as their interconnections into networks, using silver-based dyes that infiltrated the cells, making their microanatomy more thoroughly visible through microscopes than ever before.

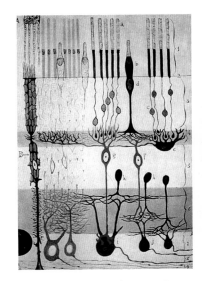

A century ago, Spanish anatomist Ramón y Cajal portrayed the retina's interconnected layers of cells.

Just as vision most often stands out as the most important human sense, those instruments that enhance vision took center stage during the early phases of the scientific revolution. But instrument makers were inventing new ways to take all of the senses beyond their limitations. The invention in 1643 of the barometer by Evangelista Torricelli, an associate of Galileo, enabled scientists to measure the pressure exerted by

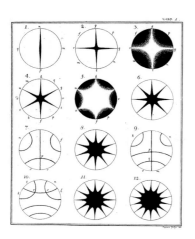

Chladni diagrams, Ernst Chladni (Discoveries Concerning the Theory of Sound, 1787).

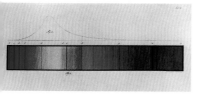

An early spectrum of the sun, hand painted by Joseph Fraunhofer in 1820.

the atmosphere. This new data was not nearly as vivid as that seen through lenses, but it was, in effect, a magnifier of the sense of touch. The resulting data helped scientists envision the sky as a vertical sea of air and, later, to divine some of the factors behind weather and climatic conditions. In the eighteenth century, the amateur musician Ernst Chladni, trained as a lawyer at the University of Leipzig, earned the title of Father of Acoustics by making sound visible. He sprinkled sand on plates made of various materials and then drew his violin bow along the edges of the plates. The sand redistributed into tantalizing patterns that revealed the plates' vibrational modes.

At about the same time, and into the nineteenth century, the pioneers of chemistry developed all manner of techniques for overcoming the primary empirical challenge of chemistry, which Cornell University chemist and Nobel laureate Roald Hoffmann says is "to know without seeing." Standing out among those innovations is spectroscopy, a family of techniques for teasing apart the various colored and/or invisible components of light. Newton had observed that white light, as it passes through a prism, spans out into a rainbow of visible colors. However, he apparently did not notice dark lines punctuating the spectrum of colors. In 1814, while testing some of his own prisms in sunlight, a glassmaker and optician in Munich, Joseph Fraunhofer, did notice the lines (due, unknown to him, to absorption of specific wavelengths of light by different chemical elements), counted up to 600 of them, and noted that they always appeared in the same locations along the spectrum. In 1820, he painted the line-riddled spectrum of sunlight along with a graph recording the relative intensities of the colors.

Not until 1859, just as Charles Darwin's theory of evolution by natural selection was beginning to transform biological thinking, did the value of spectroscopy as a means to discern the previously indiscernible begin to become apparent. The German physicist Gustav Kirchoff found that the light emanating from chemical elements that he had heated to incandescence also would disperse into a spectrum as it passed through a prism. And not only that: Each element produced a spectrum with its own characteristic, and thereby diagnostic, set of dark or bright lines, corresponding respectively to absorption and emission spectra. Suddenly, spectroscopy had become a tool of chemical analysis. Perhaps most amazing at the time, Kirchoff used the technique to infer that the sun's chemical composition included sodium. Until then, scientists had assumed that knowing the compositional studies of stars, including the sun, would be impossible.

The same challenge—to know without seeing—confronted those captivated by the phenomena of magnetism and electricity, whose common physical bases were first mathematically codified by James Clerk Maxwell in his four famous equations describing electromagnetism in 1864. But his exquisite mathematical distillation was only possible following decades of painstaking observations using tools and techniques that rendered electricity and magnetism visible, most often with clockwork-like mechanisms of a precision and delicacy that converted magnetic and electric forces into visual mechanical motions, which could be directly measured and quantified.

In some cases, scientists found ways of making these unseen phenomena visible in more concrete ways. In the early 1850s, for example, Michael Faraday capitalized on an arrestingly simple way of visualizing the magnetic field surrounding magnets. When he sprinkled iron filings around bar magnets, magnetic disks, and the individual poles of rod-shaped magnets, the filings automatically assumed patterns corresponding to what he termed "lines of force." They provided a visual schematic for formerly invisible magnetic actions, such as the way a magnet deflects a compass needle and can induce electrical currents in conductors with which it is not in direct contact.

The nineteenth century also was the century in which the invisible extensions of light and heat were revealed. In 1800, using a thermometer, William Herschel found that the invisible light emanating beyond the red side of a spectrum produced by passing sunlight through a prism produced more heat than the visible colors. He called it infrared light. A year later, the German physicist Johann Ritter discovered ultraviolet light—invisible light with colors beyond the violet extreme of the visible spectrum— when he noticed that silver chloride, a salt that later in the century would become the chemical basis of photography, decomposed more rapidly when exposed to light beyond the violet he could see.

These discoveries merely opened doors to what would turn out to be vast expanses of electromagnetic wavelengths extending beyond both the infrared and ultraviolet. In the late 1880s, Heinrich Hertz conducted his world-changing experiments showing that rapidly oscillating electric fields could create "electric waves" that invisibly propagated through the air—the principle behind radio technology. The discovery of radio waves would extend the red side of the spectrum indefinitely toward longer wavelengths.

Without quite knowing it, Wilhelm Roentgen did the same for the violet and more violent side of the spectrum when he discovered X rays in 1895. For years, nineteenth-century scientists using cathode-ray tubes—much like the ones in today's televisions— had been unwittingly creating X rays, which are not directly visible. Roentgen was the lucky one to notice them first, during a series of experiments on materials that would fluoresce when exposed to the rays of a cathode tube. On November 8, he was astonished to notice that a sheet with a fluorescent coating was glowing even though his cathode-ray tube happened to be covered with black paper, which was known to block the cathode rays. Some other kind of rays then—he soon called them X rays—must have passed through the paper shield to cause the glow.

In subsequent experiments, he found that the X rays passed through all kinds of opaque materials. The rays even passed through the skin and bones of his wife's hand, but not her wedding ring, which showed up on perhaps the first X-ray photograph of human anatomy as though it were circling a finger of a ghostly hand. The following year, after Roentgen publicly announced the discovery, doctors began using X rays to peer through their patients' skin.

Scientists' hunger to see more and more of the world in ever-more dimensions, with ever-more sensitivity, only increased throughout the last century. Within twenty years of Roentgen's discovery, the father-son team of William Henry and William Lawrence Bragg were using X rays to precisely measure the exquisitely tiny distances between the rows and columns defined by atoms in a crystal. When shined at crystals, X rays would bounce off the specimens' regimented crystal planes onto film where they would leave behind visual patterns of dots. From this new kind of data, almost a visual metaphor for

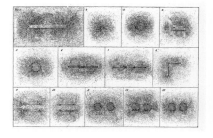

When Michael Faraday sprinkled iron filings around magnets in the mid-nineteenth century, he rendered visible the magnets' normally invisible magnetic field lines.

crystals on their finest atomic and molecular dimension, physicists and chemists could infer the precise arrangement of atoms in crystals.

As was so often the case, this new technique and its accompanying instrumentation became a window onto vast new territories of phenomena. "Thanks to the method the Braggs, father and son, have devised for investigating crystal structures, an entirely new world has been opened," said the chairman of the Nobel Committee for Physics in 1915 when the Braggs, who were unable to attend the Copenhagen ceremony due to war conditions, received their Nobel Prize in absentia. Almost four decades later, in 1952, Rosalind Franklin would gather X-ray crystallography data of DNA that would prove pivotal the following year when James Watson and Francis Crick reported that they had deciphered the structure of this most miraculous of molecules.

Similarly expansive language to describe the new vistas opened up by novel scientific instruments was heard over and over throughout the last century. In his Nobel Lecture in 1974, the Belgian-born biologist Albert Claude, who won the Prize in Physiology or Medicine for his work using electron microscopy and fractionation techniques to separate different cellular components to help uncover the complex world within cells, put it this way: "No doubt, man will continue to weigh and to measure, watch himself grow, and his universe around him and with him, according to the ever-growing powers of his tools."

Ernst Ruska, the man most responsible for developing the electron microscope that enabled Claude to visualize cellular anatomy, was himself honored in 1986 with a Nobel Prize in Physics. Because optical microscopes rely on visible light, they are unable to resolve objects much smaller than a wavelength of visible light, or a few microns, perhaps one-twentieth the width of a typical human hair. Ruska developed a magnetic lensing system that enabled him to harness electrons instead of light as a source of illumination, a shift that ultimately would lead to electron microscopes with nearly atomic resolution.

The first commercial electron microscopes went on sale in 1939—not long after Ruska had used a homemade one to obtain the first image of a virus—and this amazing tool has been undergoing a fantastic evolution ever since. The scanning electron microscope, which shoots a beam of electrons at a sample and then records the secondary electrons emitted by the sample, serves up especially arresting images: they portray a sense of depth that makes viewers feel like they suddenly have taken on sub-Lilliputian dimensions and are directly witnessing the world from this diminutive point of view. Electron microscopes have become as central to physics, materials science, and engineering as they are to biology and medicine. In its own summary of the significance of the electron microscope in 1986 when Ruska received his Nobel Prize, the Royal Swedish Academy assessed the tool's significance this way: "It is one of the most important inventions of the century."

The proliferation of instruments for making the invisible visible continued unabated throughout the twentieth century. The loss of the *Titanic* as well as wartime concerns drove the development of sonar—once known in long form as "sound, navigation, and ranging"—to enable mariners to detect dangerous and otherwise-unseen objects, including submarines and the parts of icebergs that are below the water's surface. The technique works by sending out pulses of sound and detecting their echoes. Sonar relies mostly on low-frequency sound, but its much higher-frequency sibling, ultrasonic imaging, has been put to use for viewing inside things without having to cut into them. For years now, ultrasonic portraits of fetuses have preceded birth pictures in family albums.

The development of radar—once known as radio detection and ranging—before, during, and after World War II has led to fantastic sort of distance vision, in which people can detect and track objects thousands of miles away. Because radar is based on radio waves, which can penetrate clouds and are indifferent to lighting conditions, radar-imaging devices are effective in all kinds of weather and at all times of day. Many of the most stunning images of Earth's surface, as well as the surfaces of other planets and moons, have come from airborne and space-based radar instruments.

One of the most consequential inventions in the past few decades is an instrument known as a scanning tunneling microscope (STM) that, in a sense, extends our sense of touch to the point where we can feel atoms. Using STMs and the diversity of so-called scanning probe microscopes (SPM) that they have spawned, scientists have been able to visualize the atomic and molecular landscapes of everything from silicon chips to biological cells. The crux of an STM is a superfine stylus—sometimes atomically fine— that scans back and forth over a surface like a tiny finger reading Braille. The stylus is scanned so close to the sample that electrons in the tip jump the gap even though there is no conductor to carry them across, in a quantum mechanical phenomenon called tunneling. The strength of this tunneling is a direct reflection of the distance between the tip and the surface. Using an ingenious mechanism that slightly lowers and raises the scanning tip in order to maintain a constant tunneling current, a computer can gather enough topographic detail about the sample surface to reconstruct an image.

So swiftly was the STM embraced by the scientific community that only five years after unveiling it its inventors, Gerd Binnig and Heinrich Rohrer of IBM Research Laboratory in Zurich, Switzerland, shared the 1986 Nobel Prize in physics with Ruska. The STM and its progeny—including the atomic force microscope which, unlike the STM, can produce images of nonconducting samples—have changed the look of data in many scientific fields. Less than a century after the very existence of atoms was a topic of debate among the world's greatest scientific minds, SPM images of atoms and molecules have become routine.

IN COMING years, sophisticated scientific instrumentation will bring more and more previously invisible phenomena out into the open for human eyes to finally see.

Consider one of the largest scientific instruments ever built—the relativistic heavy ion collider (RHIC) at Brookhaven National Laboratory on Long Island. With it, scientists can accelerate gold ions to 99.95% the speed of light and then smash them together in the most energetically concentrated head-on collisions ever produced on Earth. The goal is for the protons and neutrons comprising the ions' nuclei to melt back into quark-gluon plasma—the primordial stuff from which the constituents of normal matter presumably froze out once the infant universe had cooled down enough. Cosmological theory holds that quark-gluon plasma existed for only tens of millionths of a second after the Big Bang, though it may still exist out there in the cores of unimaginably dense stars known as neutron stars. If all goes well, the RHIC will provide portraits of this moment when protons and neutrons first formed. For all of its grandeur, it is a tool for looking in on the microscale of matter when matter itself was a new thing.

In the works are instruments for scanning the universe on macroscales as well. One of them is the James Webb Space Telescope (JWST), the successor to the Hubble Space Telescope, slated for launch in 2009. The Hubble's mirror and detector can gather light from galaxies up to about five billion light-years away, which means it provides astronomers and astrophysicists with large-scale views from approximately the last third of the universe's existence. Light from older galaxies and other cosmic structures is shifted into the infrared portion of the spectrum, so the JWST will be a thousand times more sensitive than existing and other planned observatories and presumably will capture images of the universe's first generation of stars and galaxies as they formed from the primordial nebulae of mostly hydrogen gas.

To complement these large views, designers now are working on a host of instruments for mapping the surfaces of nearby stars, rather than merely analyzing their electro-magnetic emissions; for studying the stellar populations of other galaxies with the kind of clarity that now is only possible in our own Milky Way; and for directly imaging planets that may be circling any of the thousand stars closest to our own.

On NASA's drawing board, for example, is an instrument dubbed the Planet Imager, designed to image extrasolar planets not as single dots near their parent star, but as more detailed depictions that might reveal whether the planets have oceans, continents, and other features. One design for an instrument with this capability includes an array of five space-born interferometers—each one harboring four 8-meter telescopes for gathering light—that would fly, according to those envisioning it, in "exquisitely precise formation…over distances of 6,000 kilometers or more!" Such a system is expected to obtain 25 by 25 pixel images of planets around nearby stars with the kind of detail illustrated in the second most forward image in the quartet shown here. This has been described as like standing in New York and being able to see a fly buzzing by a spotlight in Hollywood. The foreground Earth is depicted with a 10 x 10 pixel resolution and the background Earth is depicted with a 400 x 400 pixel resolution.

While we cannot predict future scientific breakthroughs or the instruments that will make them possible, it is certain that the data from these tools will be packaged in ways every bit as arresting and compelling as the images in this book. The visual landscape is destined to become enriched with ever more scientific imagery. The image-processing tools required to do so are sure to be part and parcel of the instrumentation itself; the standard accoutrements of the laboratory will encourage scientists to render their data more attractively. The ubiquity of such presentation tools, along with high-tech digital projection systems that fill large screens with crisp high-quality imagery already is too much for scientists to resist. When they tell their stories and show their

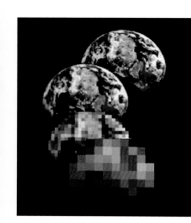

Future telescopes may obtain images of planets around other nearby stars with the resolution equivalent to that of the Earth image second from the bottom.

data in the future, investigators will want to dazzle their audiences with data of increasing grandeur and beauty, with data that is indistinguishable from art because it has, in fact, become an art form unto itself.

Lynn Boatner, a materials scientist at Oak Ridge National Laboratory in Tennessee, already has adopted this attitude. It is impossible to view his microscope images of metallurgical phenomena without imagining them on the walls of an art gallery. Says Boatner: "Nature exhibits both symmetry and chaos, and from time to time either or both of these characteristics can lead to images on every dimensional scale whose value transcends the scientific interpretation of physical phenomena and extends into the realms of art and esthetics."

The growing body of "data art" is more than just a delightful bonus for scientists; it's feeding back into the process of science. "Attempting to render physical phenomena carefully, even artfully, can lead to scientific discovery," says Eric Heller, a Harvard University physicist whose stunning computer simulations of complex phenomena, such as electrons flowing in semiconductors, are, in his words, "teaching us what happens in nature." For Heller, the imagery emerging from his simulation studies not only reflects the physical models and experimental data that underlie the simulations, but it helps him and other scientists to discern real-world phenomena that might otherwise remain unnoticed.

Science will continue to open up spectacular new views of nature, views that are sure to open up pathways to new scientific insight. It is a precious cycle that gives to us a Super Vision.

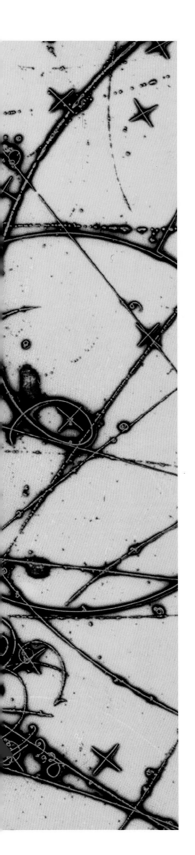

Particularity Toward the end of Long Island, physicists at the Brookhaven National Laboratory have been smashing atoms together for decades. Their aim has been to study the fundamental particles that comprise atoms and, thereby, just about every object ever encountered. In the 1950s and 1960s, one favorite way of rendering visible the subatomic debris caused by these collisions was to stage the crashes inside sometimes room-sized tanks of liquid hydrogen known as bubble chambers. In such chambers, particles emerging from the collisions leave tracks of tiny bubbles, which were photographed in black and white and then occasionally colored using computer graphics tools, as in this case (above). Magnetic fields infiltrating the chamber caused the trajectories of charged particles to bend one way or another, and sometimes into spirals, while uncharged particles careened through like bullets. Although the streaks of bubbles were big enough to photograph, the particles that caused them were of dimensions similar to atomic nuclei, or roughly one-millionth of one-billionth of a meter (10^{-15} meter).

Subatomic Sightings As particle accelerators became bigger and more powerful over the second half of the twentieth century, so did the detectors for recording collisions between the high-energy particles. One of these detectors was known as the Big European Bubble Chamber (BEBC) and was located at CERN, the European Center for Nuclear Research, in Geneva, Switzerland. The garage-sized BEBC contained so much flammable liquid hydrogen that the laboratory had to maintain a hotline to the local airport as part of a warning system to be used in case of a sudden release of hydrogen into the air. During its working lifetime, from 1973 to 1984, the BEBC yielded more than six million pictures of particle collisions. Here, a superconducting magnet girdling the chamber coerced the trajectories of charged particles emerging from otherwise invisible collisions into curved pathways. By measuring the curvatures, physicists could calculate how much energy the particles carried and what kind of particles they were. Depicting a morass of particle cascades due to multiple collisions within the chamber, the image at left is particularly striking, made even more so by the artificially colored blue of the collisions' tracks and the amber of the background. The subatomic particles causing the collisions typically had diameters in the range of an atomic nucleus, about one-millionth of one-billionth of a meter (10^{-15} meter).

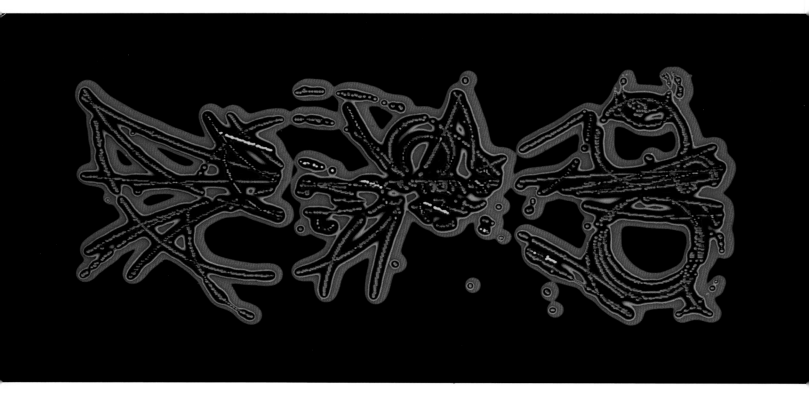

Creative Carnage One implication of Einstein's famous equation, $E = mc^2$, is that matter can transform into energy and energy can, in effect, congeal into matter. This image from the 1980s depicts these processes, brought about by colliding a positively charged proton, which is about two thousand times heavier than an electron, with an antiproton, which is basically a negatively charged proton. The proton and antiproton annihilated one another. The energy of that process produced a barrage of particles, whose resulting tracks are shown as blue lines within red halos. Among these particles was a so-called Z particle, which instantly decayed into an electron and its positively charged antiparticle, a positron. The trajectories of the electron and positron are shown as yellow lines emerging in opposite directions from the center. Analysis of the almost calligraphic assemblage of lines provided the smoking gun for physicists who were in hot pursuit of the proposed Z particle in the 1980s. This particle is the carrier of the weak force, a fundamental force governing the structure and behavior of atomic nuclei. The particular smash-up depicted here was recorded in a detector monitoring collisions orchestrated by the Super Proton Synchrotron—a particle accelerator and collider nearly four miles in diameter that is part of the European Center for Nuclear Research in Geneva, or CERN. A proton's diameter is on the order of one-millionth of one-billionth of a meter (10^{-15} meter).

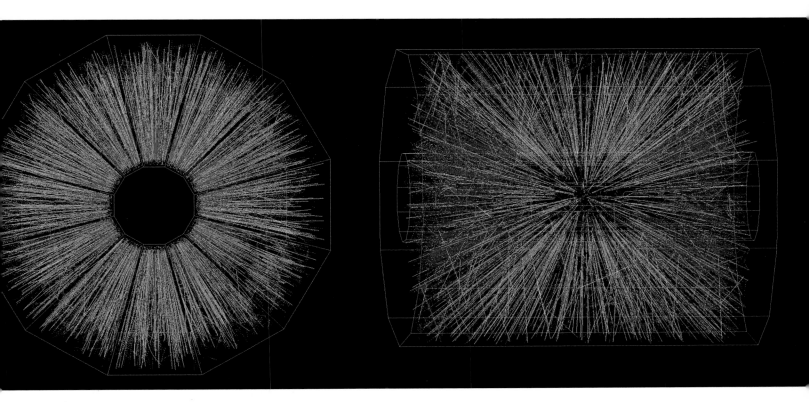

Searching for the Mother of Stuff Before there were any galaxies or suns or planets or people, before there were even any atoms, there was ur-stuff, a primordial form of matter-in-potential that physicists refer to as the quark-gluon plasma (QGP). From this soup of fundamental particles—specifically quarks and gluons —atoms froze out, making possible the evolution of our familiar world of objects. In an attempt to recreate and study a tiny dot of QGP, physicists have built colliders that can smash some of the periodic table's larger atoms together with enough energy so that the atoms' constituents possibly could melt back into QGP droplets. On June 12, 2000, at 9 PM, two gold ions (gold atoms stripped of their electrons), smashed head-on at the Relativistic Heavy-Ion Collider (RHIC), which is housed at the Brookhaven National Laboratory on Long Island. The ions were traveling at about 99.9 percent the speed of light. Although there were no signs of QGP from the collision, the resulting cascade of particles, depicted here from two different perspectives, provided graphic evidence that the RHIC ultimately could energize ions enough to produce the coveted QGP. The gold ions have diameters of roughly one-hundredth of a trillionth of a meter (10^{-14} meter).

Anatomy of Reactions The process by which molecules react and undergo change has largely been hidden from view. Molecules are too small to see, and reactions often happen in billionths of a second or less. Nonetheless, scientists have been finding ways to visualize reactions. Here, chemists used a laser to stimulate the molecules in a beam of nitrogen dioxide (ONO) to split into two fragments—an oxygen (O) fragment and a nitrous oxide (NO) fragment. The fragments can be made to absorb light from a second laser, which ionizes them. The investigators were able to visualize the NO fragments on the equivalent of a TV screen as they were ejected from the parent nitrogen dioxide molecules. In the image, they appear collectively as mushroom-shaped fronts zipping up and down. From the overall distribution of the ejecting fragments, the scientists could infer that the laser-induced fragmentation occurred in much less time than it takes the individual nitrogen dioxide molecules to rotate within the beam. Since they already knew that rotation rate, they could use it to measure how long it took the ONO fragmentation reactions to occur. Understanding these kinds of details about reactions can, for example, help chemical engineers develop more efficient industrial processes and can provide clues to those who might want to use beams of laser light for controlling chemical reactions to make new materials, or, possibly, quantum computers. The diameter of the beam in which the fragmentation occurs is about 1.5 millimeters (1.5×10^{-3} meter), and the chemical bonds between the nitrogen atom and each of the two oxygen atoms in a nitrogen dioxide molecule span about 1.188 angstroms (1.188×10^{-10} meter).

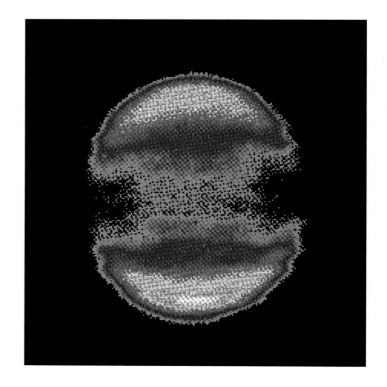

Colorizing Genes An early step in the gene-sequencing process is to rip an organism's genome into small pieces, akin to tearing a book into individual pages or even paragraphs. The resulting DNA snippets then often get inserted into bacteria, which make copies of the snippets simply by virtue of the microbe's own tendency to replicate itself. The replicated snippets undergo another series of molecular "photocopying" steps using a laboratory technique known as polymerase chain reaction (PCR), which yields enough copies of each snippet for a robotic sequencer to read its sequence of DNA nucleotides. Each of the four nucleotides is marked with a particular dye, and a laser system detects the dye of each nucleotide as it comes through the analyzer and creates a four-color ribbon of data corresponding to the DNA's nucleotide sequence. The sequence data shown here is from the fungus *Cryptococcus neoformans*, which is known to cause meningitis in susceptible hosts, such as those with AIDS. Each lane of colors reflects the sequence of a specific snippet of DNA. The spacing between nucleotides along a DNA molecule is about 3.4 angstroms (3.4×10^{-10} meter).

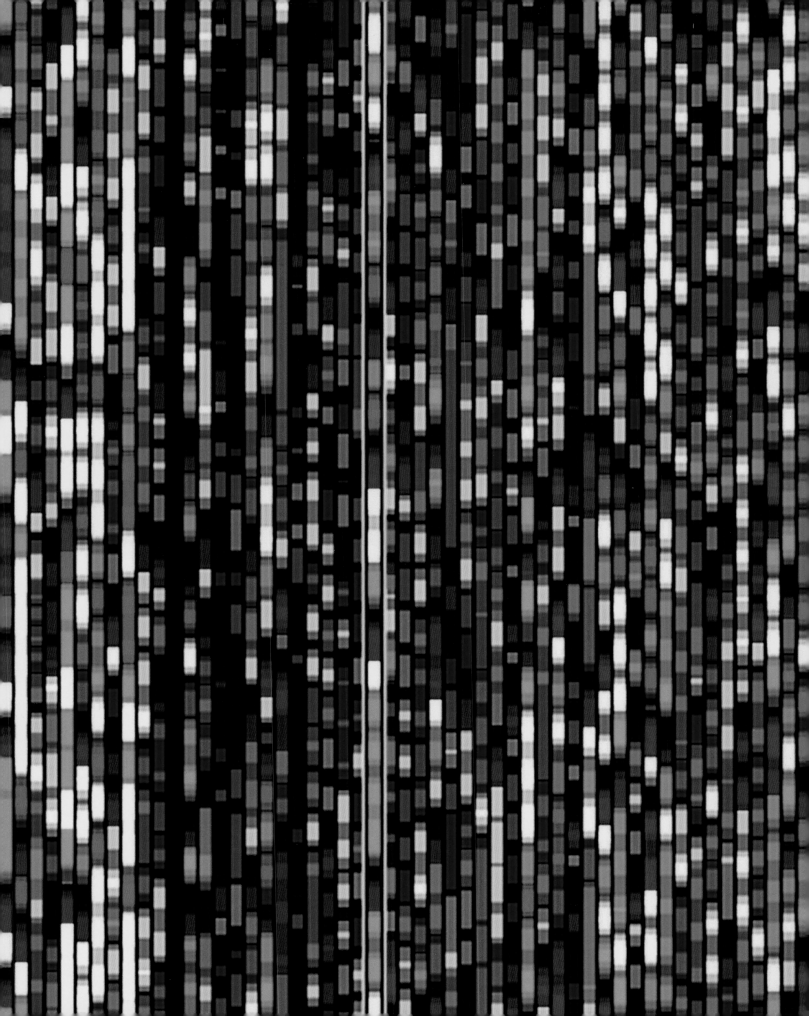

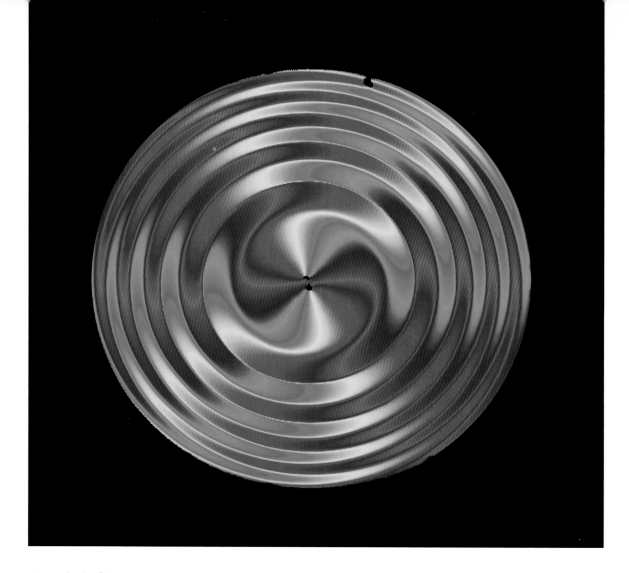

Crystal Birefringence Crystals are ordered structures whose components are stacked in regimented layers. In most cases, this means light passing through a crystal will experience a different environment depending on the light's polarization, essentially the orientation the light waves have as they propagate through space. These different environments make the crystal birefringent; light traveling through it will bend and appear differently colored depending on the direction from which the light enters. Quartz is an unusual material in that it shows two kinds of birefringence, linear and circular. By imaging a tiny cone of light passing through a quartz specimen using a birefringence-imaging microscope, scientists can study both types of birefringence at once, revealing several properties of the material. The concentric rings in the image at right provide a measurement of the crystal's linear birefringence, the property that makes light bend to different degrees depending on its incoming direction. The image above, a so-called orientation image, reveals that light traveling through this quartz crystal undergoes a leftward (levorotatory) rotation. These properties derive from the smallest structural scales of the quartz crystal, the scale of atoms and molecules, which is in the range of several ten-billionths of a meter ($\sim$$10^{-10}$ meter).

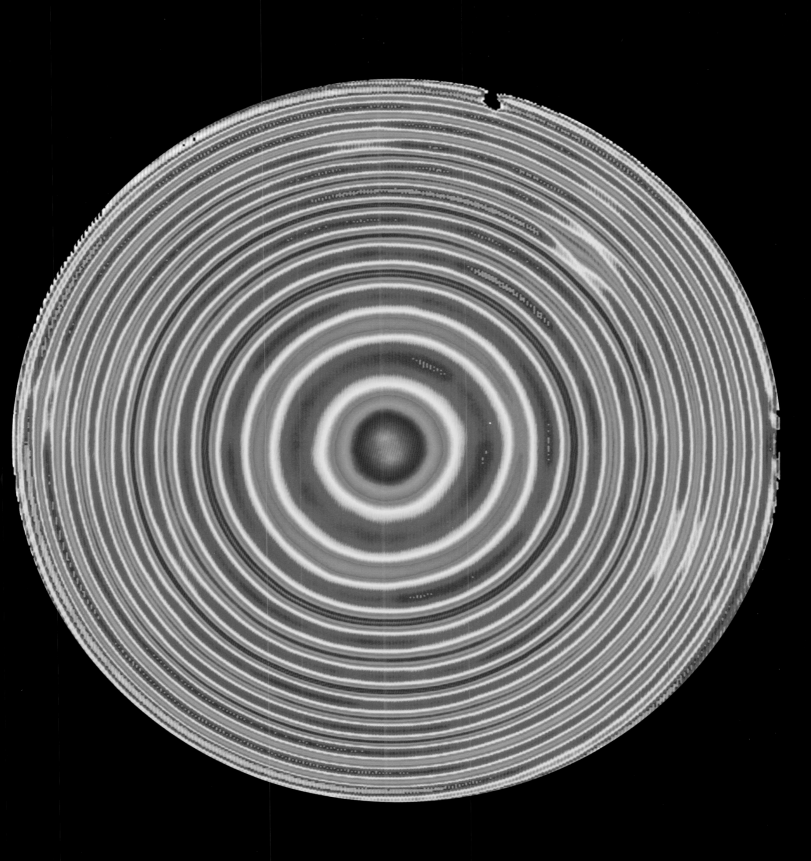

Crystal Mandala It is the details, on exquisitely fine levels, that determine whether any particular arrangement of atoms is an electrical conductor or an insulator, whether it's hard or soft, brittle or tough. Materials
researchers at the University of Cambridge in England beamed electrons into a sample of lanthanum
aluminate ($LaAlO_3$), a ceramic substance that in the 1980s helped rekindle research in superconductivity. The
scientists measured two kinds of electrons diffracting from the aluminate in a telling spray of angles. Some
of those electrons reached detectors in patterns that formed the concentric disks and the lines radiating from
the core of the image. Working with these features of the "mandala," whose pleasing colors were added later,
the Cambridge scientists could discern the overall crystal architecture of this substance. They were able to
determine the shape of the crystal's basic unit—its so-called unit crystal—as well as calculate the space
between the units. The more detailed features within the mandala's disks enable the researchers to pinpoint,
with subatomic precision, the positions of the material's lanthanum, aluminum, and oxygen atoms within
each unit crystal. With such information, the atomic bases of a material's properties are more readily
identified. The data also help researchers to improve existing materials or design entirely new ones, perhaps
even substances that are harder than diamond or better superconductors, which carry electricity resistance-
free. The spacing between unit crystals of lanthanum is 5.357 angstroms (5.357×10^{-10} meter).

Finding Microgems It is often the small stuff in a material that makes all the difference. A wafer of absolutely pure silicon would be about as useful as a paperweight. Spice the wafer with atoms like phosphorus and nitrogen, however, and suddenly the crystal has stepping-stones for electric charges, and it becomes the famous semiconductor that has changed the world. Techniques that can reveal such subtle differences on tiny scales have become crucial in modern technology. In this case, researchers at Sandia National Laboratory in New Mexico exploited a technique known as electron backscatter diffraction. Incoming electrons scatter backward in slightly different ways from different materials due to the specific and characteristic spacings between the atoms of those different materials. This image, which dates to about 1994, proved the principle of the technique using a crystal of wulfenite (lead molybdenum oxide). The dominant star-like pattern reveals—to those initiated in the art of reading these patterns—that the wulfenite crystal structure resembles a shoebox with one end shaped like a square. The four-armed star in the upper part of the image reveals that particular feature. The other lines in the sample reflect the many internal planes of the wulfenite crystal, yielding enough data for researchers to determine a detailed picture of the material's atomic-scale structure. Adjacent crystal planes in wulfenite are separated by 5.435 angstroms (5.435×10^{-10} meter).

Atomic Liaisons With new imaging tools, including the scanning tunneling microscope (STM), which began making it into laboratories in the early 1980s, scientists can both visualize and manipulate the physical world on the atomic level. Researchers used an STM to capture two images that were subsequently superimposed to answer a basic question about how a metal surface and an atom of gas might interact: Where does a xenon atom settle onto an underlying surface of nickel atoms? STM images are reconstructions of the trajectory of the instrument's tip over a sample, which entails that computers are heavily involved. The grid of magenta drop-like spots represents the tops of nickel atoms detected with an STM. Superimposed on top of that image is a second one that is almost the same, but includes a single xenon atom—shown in light blue—that has situated itself in a specific location. To see if this atom rests between nickel atoms or directly on the crest of a nickel atom, an STM innovator digitally "sliced off" the top of the xenon atom, looked inside, and found that it perched on a nickel atom like one marble on top of another. The diameter of a xenon atom is nearly 5 angstroms, or .5 nanometers (5×10^{-10} meter).

Lifeline Viruses are life at its barest and simplest. Hardly more than diminutive containers of gene-encoding molecules, either DNA or RNA, they rely on the molecular machinery of other organisms to procreate. Shown here, in a high-magnification electron micrograph, is the DNA of the bacteriophage (a kind of bacteria-attacking virus) known as lambda. Electron microscopy works roughly like optical microscopy, only it is electrons instead of light that "illuminates" the samples. The electrons are smaller than the wavelengths of optical light, so the technique can image extremely tiny objects, sometimes even molecules. The DNA was made visible with a shadowing technique—it was attached to a slide and then coated with platinum, which interacts strongly with electrons. A lambda virus had injected its own DNA into an *E. coli* bacterium, which then began making copies of the virus' genetic material. The ill-fated host produced several hundred copies of the phage until, after about twenty minutes, the host's cell wall broke apart and the new phages emerged in search of their very own bacterial hosts. The DNA molecule shown here was extracted from the bacterium at an intermediate point during the replication process. It was also chemically "denatured," or straightened out from its ordinary clumped conformation. Unfurling the DNA made the chain-like molecule more visible and helped researchers find the point or points along the molecule where replication began. In several places— where the molecular line bifurcates for a spell and then comes back together—the relationship between molecular template and molecular copy becomes visible. The DNA molecule is about 2.5 billionths of a meter (2.5 x 10^{-9} meter) thick and perhaps, in this outstretched form, about a thousand times longer (~10^{-6} meter).

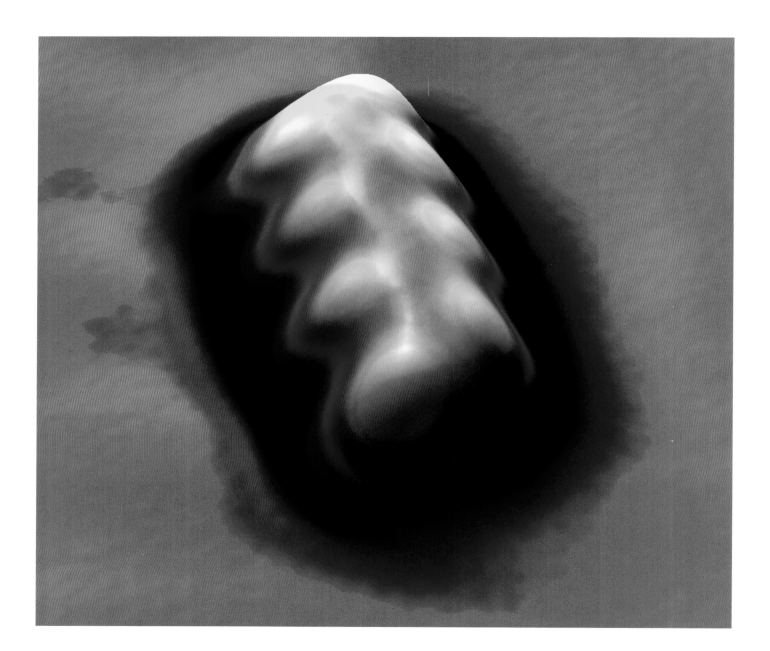

Handmade Molecule Using miraculously powerful tools, researchers are able not only to discern individual atoms on surfaces, but manipulate them as well. This molecule was virtually handmade by scientists using a scanning tunneling microscope to grab and drag individual atoms to precise locations. In this case, they dragged eight cesium atoms and eight iodine atoms over the corrugated surface of a copper crystal to produce a sixteen-atom cesium iodide molecule, which they then imaged using the very same instrument they used to make it. The color and shading were added later to help highlight its shape and to make the image pretty. Although the molecule is not useful in itself, developing the skills to deliberately make molecules atom by atom could open pathways to new, more capable materials and tiny structures, such as data-storage media of almost unimaginable capacity. The cesium iodide construction is 2.8 nanometers long (2.8×10^{-9} meter).

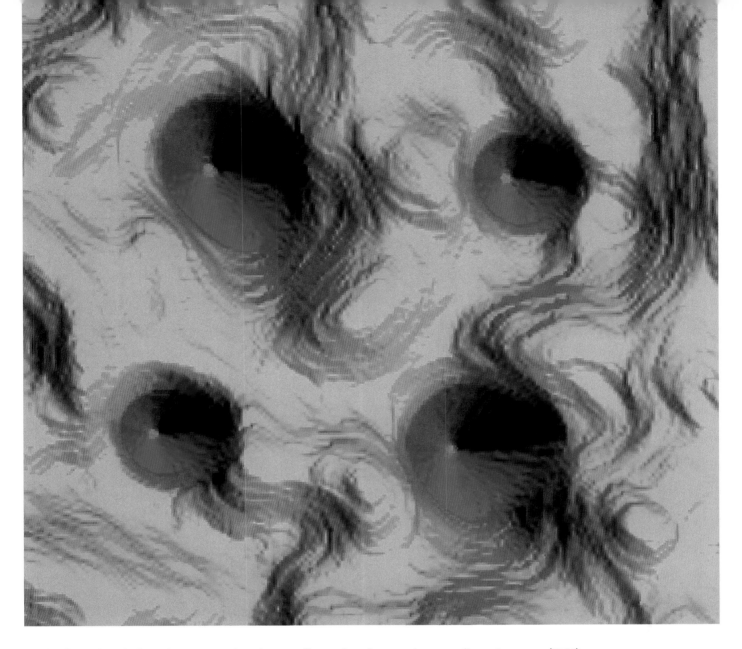

Still Life with Cobalt and Copper Using the superfine stylus of a scanning tunneling microscope (STM), Joe Stroscio of the National Institute of Standards and Technology and his colleagues did some superlatively fine handiwork. They dragged exactly six cobalt atoms over a crystalline grid of copper atoms, which appears as an orange background, to create an atomic construction and then used the same STM to visualize the result. The two smaller blue islands reveal the presence of single cobalt atoms and the larger ones correspond to pairs of cobalt atoms. The square arrangement of cobalt atoms acts as a magnetic impurity that restricts the motions of electrons in the copper surface in such a way that the wave nature of electrons—a quantum mechanical phenomenon only recently made visible by such instruments as the STM—becomes discernible as squiggly lines, in this case colored red and green. To render this quantum behavior visible, the NIST researchers, who are among a growing army of scientists and engineers who are developing tools and techniques for building structures atom by atom, had to chill their tiny structure to a mere 2.3° K, or just above absolute zero and certainly colder than the surface of Pluto. The area of the image is roughly 8 nanometers on a side (8 x 10^{-9} meter).

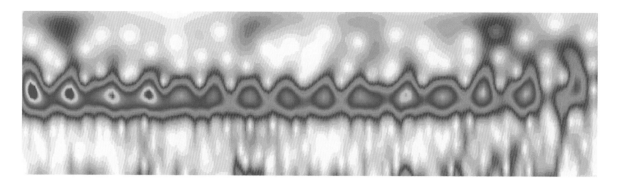

A Telling Beat As you sit in a parked car, your heart beats. Each beat generates a small shock wave in your body. It doesn't stop there. The shock wave spreads throughout the car by way of the seat, door, steering wheel, and any other solid part of the car you may be in contact with. Using a sensitive microphone placed on the car, these shock waves can be detected and then displayed as a human ballistocardiogram, kin to the electrocardiogram used by the medical community. This technology could become the basis of a surveillance technique well suited for detecting people hiding in parked cars or on the other side of walls. It also could be helpful for finding people trapped in debris following an earthquake or a terrorist strike, and several prisons have tested it as a technique to guard against inmates being smuggled out. A ballistocardiogram begins when a sensitive geophone or a microwave-based motion detector senses the tiny heartbeat-caused deflections in a car, wall, or other object. Because the signal is so weak, the federal researchers who developed this technology rely on a mathematical tool known as wavelet analysis, which can spot key features in weak and otherwise difficult-to-process signals. Each heartbeat generates vibrations of perhaps a few hundredths of micron ($\sim 10^{-8}$ meter).

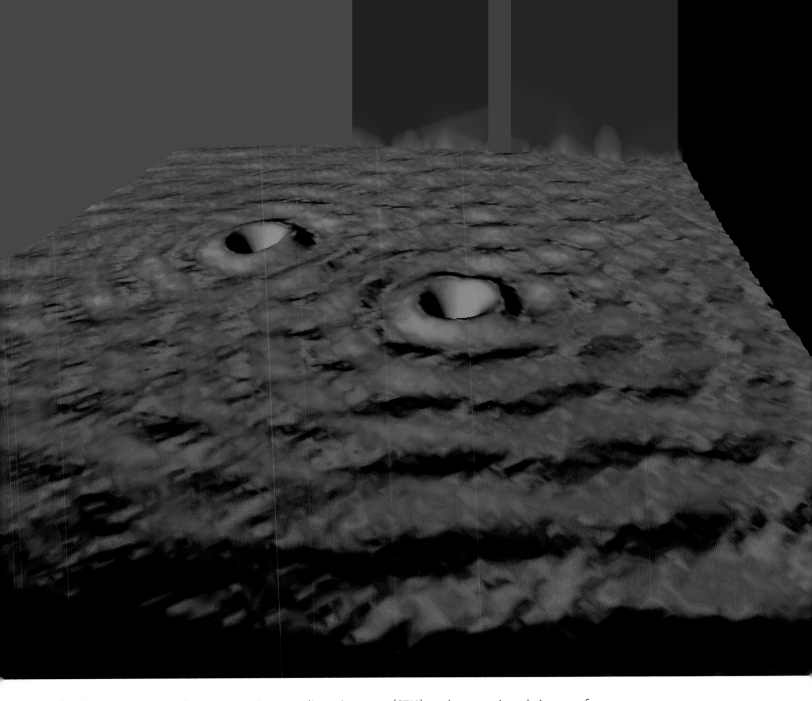

Defective Beauty Not only can a scanning tunneling microscope (STM) produce atomic-scale images of surfaces and move atoms, but it can also register the behavior of electrons moving within that surface. Here, a crystalline surface of copper atoms—shown as a subtle blue corrugation—is marred by two so-called point defects, which appear here as depressions in the copper surface and are caused by an impurity, probably sulfur atoms. Like water going by stones in a river, the electrons in the copper surface respond to these defects by restricting their "location" to a pattern reminiscent of standing waves that form in a pond when a pebble is dropped into it. The wave nature of matter was never as readily discernible as it is now with tools like the scanning tunneling microscope. The area shown is about 22 nanometers on a side (2.2×10^{-8} meter).

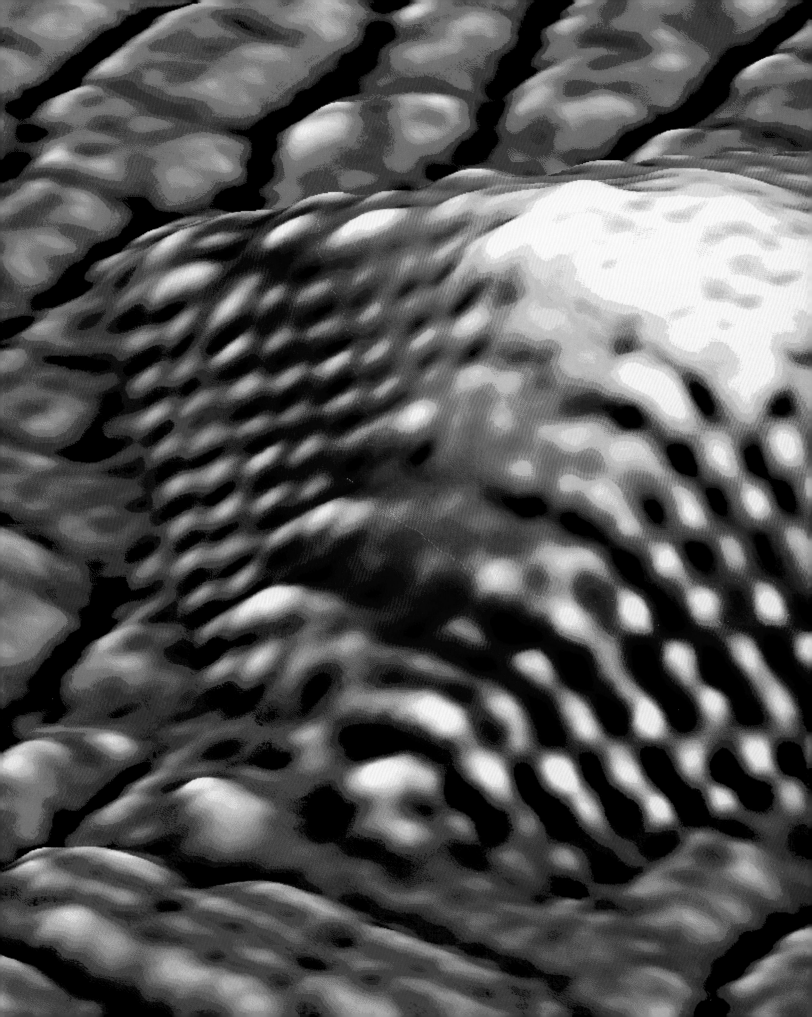

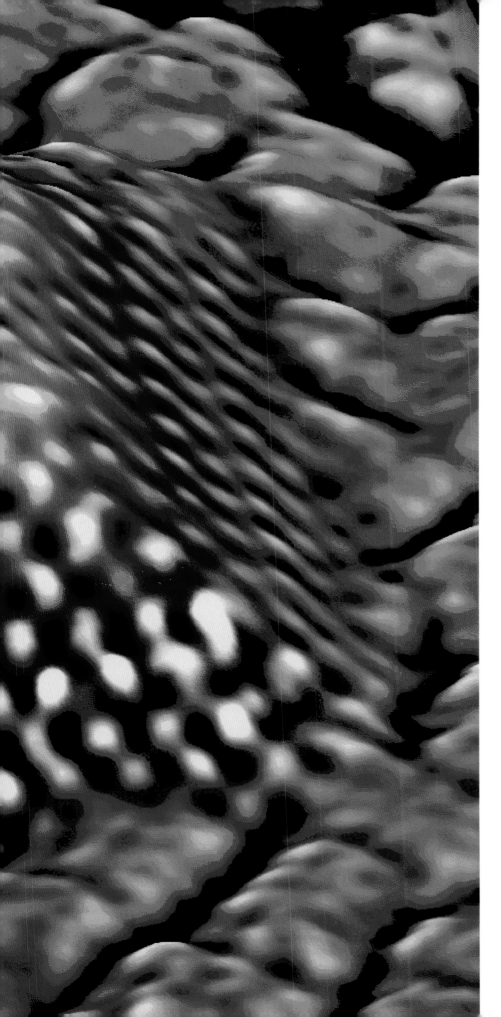

NanoCheops This nanoscale pyramid of germanium atoms—one incarnation of a quantum dot, a tiny structure that behaves somewhat like an artificial atom—formed spontaneously atop a crystalline ground of silicon. Researchers are hoping to develop quantum dots into new generations of tinier electronic devices. One application would be to use vast arrays of the minuscule dots for storing massive amounts of data in ever-smaller spaces. Also, since these structures are so small, they potentially can be exploited to build entirely new kinds of chips that may be far more powerful than conventional chips. The image was acquired with a scanning tunneling microscope, which does not record colors; colors are added either automatically by the computer associated with the STM or after the fact, using image-processing programs. Each side of the quantum dot is about 25 nanometers long, about the same as its height (2.5×10^{-8} meter).

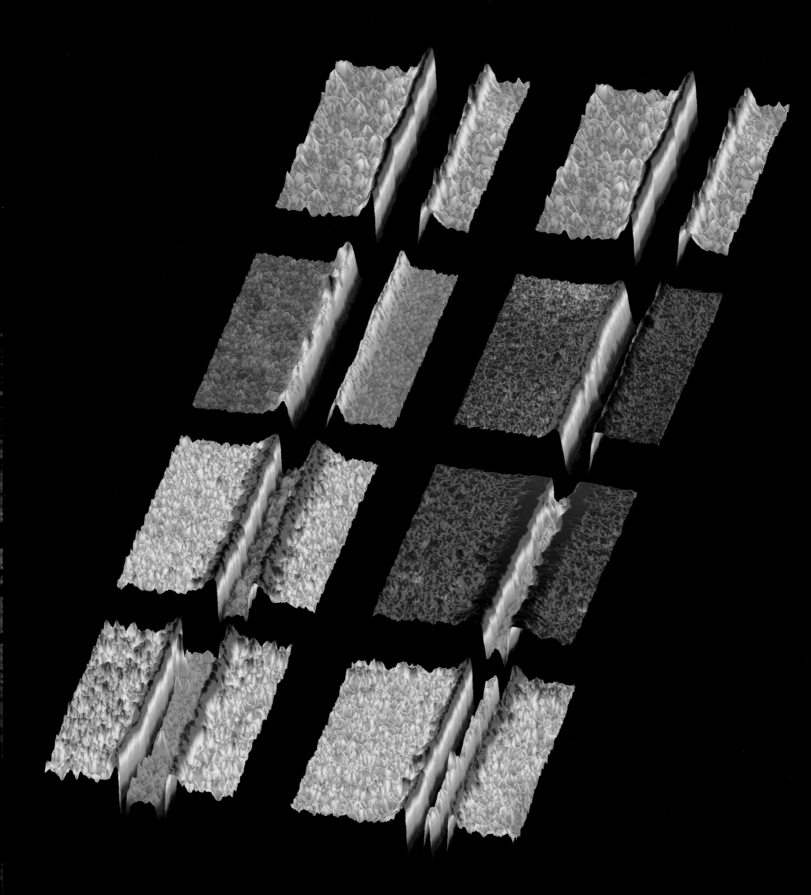

In December 1959, just as the era of technological miniaturization and microelectronics was about to accelerate, the storied physicist Richard Feynman gave a talk at the California Institute of Technology. Titled "There's Plenty of Room at the Bottom," it has become a veritable liturgical document for high technology. He spoke of such amazing prospects as storing the entire contents of the Library of Congress in a shirt pocket. Before the century gave out, scientists had indeed developed many remarkable ways of recording lots of information in small spaces. One of them even resembles the old dip pen technology that scribes used centuries ago. But the dip pen used for the writing in this image was the tip of a scanning probe microscope and the "ink" was single layers of certain organic molecules deposited on a gold surface. In a nod of deference to Feynman, scientists used the device to print a paragraph of the physicist's famous lecture. The letter "l" is only 60 nanometers wide, or about one-tenth the length of a single wavelength of yellow light. At that size, it would be possible to store a trillion bits of data—roughly the equivalent of tens of thousands of *Encyclopedia Britannicas*—on a surface smaller than a postage stamp (6×10^{-8} meter).

Fine Wire Act Though engineers may have honed the microfabrication techniques used to make high-technology gadgetry like microprocessor chips, they realize that their present strategy—which is based on transferring circuitry patterns on masks to silicon wafers by way of light, lenses, and various materials that etch away the silicon or deposit onto it—will have been pushed as far as possible within the next ten years or so. In a bid to keep the miniaturization trend going, many researchers are turning to nanotechnology, which refers to technologies that engage the world on the billionth-of-a-meter scale, or roughly the scale between individual molecules and viruses. To make the nanoscale structures depicted in this image, scientists first used highly focused beams of electrons to deposit on top of a silicon wafer two ultrathin lines of gold separated by a trench of about 110 nanometers, or about 1/1000 the width of this "t." Two of these wire structures are shown, via high-resolution scanning electron microscopy, at the top of each column. With these structures in hand, researchers then deposited ten layers of an organic "resist" on the left structure (shown in blue in the second row) and twenty such layers atop the right (shown in red). The resist layers served as high-precision masks, narrowing the exposed width of silicon in the trench to 65 nanometers and 25 nanometers, respectively (shown in blue in the third row). Some viruses would be unable to fit in these gaps. In a final step (bottom row), the researchers deposited gold into the center of the silicon trench on the right and then washed away the organic resist material. Left behind in the trenches were some of the finest wires ever made (shown as a free-standing yellow-topped wall). The wire in the bottom right trench is about 10 nanometers thick (10^{-8} meter).

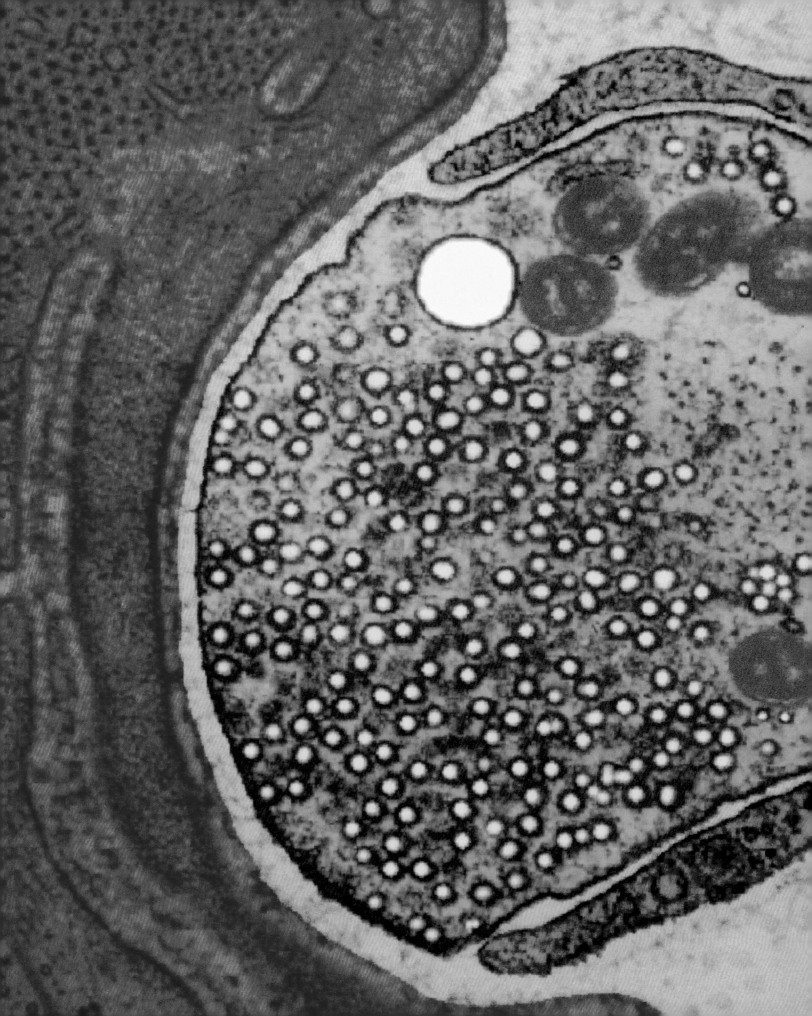

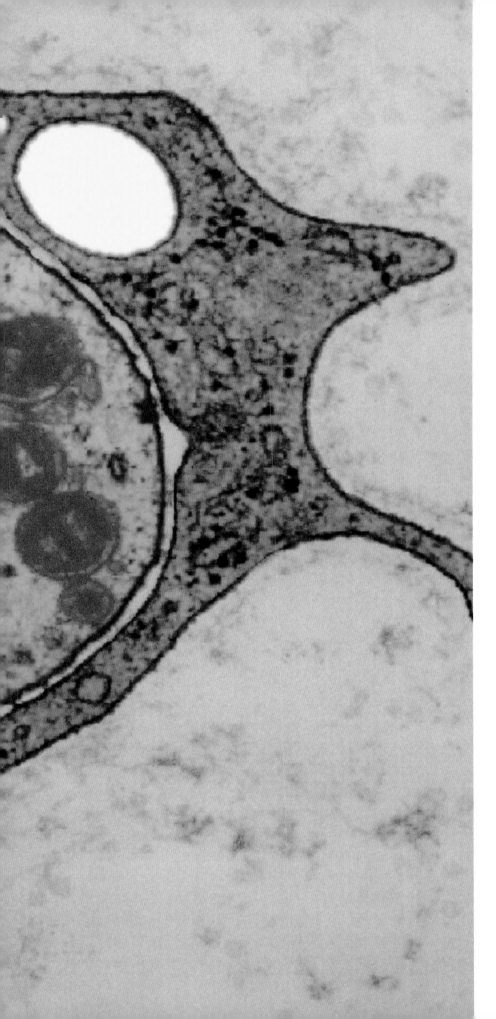

Nervous Monad One of the greatest discoveries about the brain is that it consists of trillions of cells that interconnect into a wondrous network of electrochemical activity. Neuroscientists have learned enormous amounts about the brain's structure and functions—from the global signals detectable via electroencephalogram (EEG) equipment to the microscale communication that occurs at synapses between cells and that relies on signal-relaying molecules known as neurotransmitters. In this transmission electron microscope image, electrons passing through a thin slice of tissue revealed the intimate cellular connection between a nerve cell and a muscle cell. The end of the nerve cell's signal-carrying axon, in blue, nestles into a cleft in the muscle cell, creating a neuro-muscular synapse. The green structure is part of a Schwann cell, which provides the neuron with metabolic support and helps it transmit impulses. The rust-colored donut-like structures in the nerve ending are mitochondria, the cell's tiny fuel-making factories. The smaller circles are vesicles filled with the neurotransmitter acetylcholine. When the nerve fires, some of these packets explode open, spewing acetylcholine molecules into the synaptic gap between the cells. When receptors on the muscle cell receive enough of the neurotransmitter, the cell responds by setting its contractile micromachinery into motion. When enough muscle cells do this at the same time, an entire muscle can move, bringing with it, say, a leg or an arm. Each synaptic vesicle is about 50 nanometers in diameter (5×10^{-8} meter).

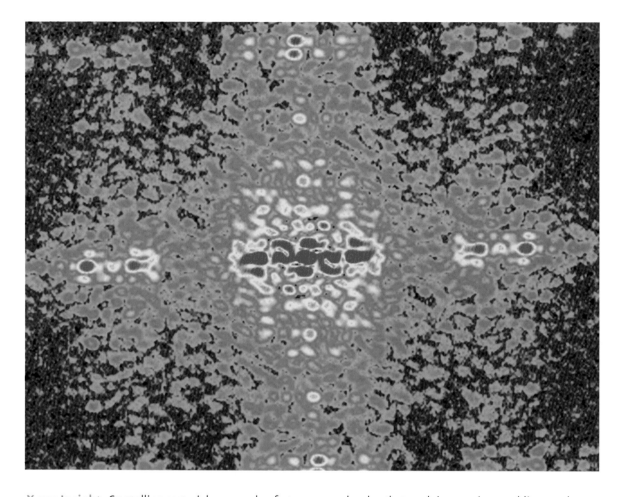

X-ray Insight Crystalline materials are made of atoms or molecules that pack into regimented lines and planes forming three-dimensional geometries, often resembling playground jungle gyms. Since the early twentieth century, researchers have used X rays to determine internal structures of crystals. The technique, known as X-ray crystallography, works because the wavelengths of X rays are short enough that they diffract —a kind of microscale reflection—from a crystal's internal atomic and molecular planes. Using mathematical tools and computers, the resulting diffraction pattern can be parsed to reconstruct the crystal structure responsible for the pattern. Because noncrystalline materials do not diffract X rays with the same kind of regularity, their internal structures have been harder to study. This image represents an early success in the late 1990s in using X rays to determine the structure of noncrystalline materials. In this case, researchers deposited virus-sized dots of gold (shown in red) onto a surface of silicon nitride, much as a dot-matrix or ink-jet printer uses tiny dots of ink to create discernible letters. Then, with X rays from the National Synchrotron Light Source at the Brookhaven National Laboratory on Long Island, the scientists produced an X-ray diffraction pattern for the noncrystalline array of gold nanodots. The researchers were able to then work backward and extract from this raw data an image of the original nanodot structure at a resolution of 65 nanometers. The method holds potential for imaging small biological cells, such as bacteria, and even large subcellular structures, such as nuclei. Each gold nanoparticle is about 100 nanometers in diameter (10^{-7} meter).

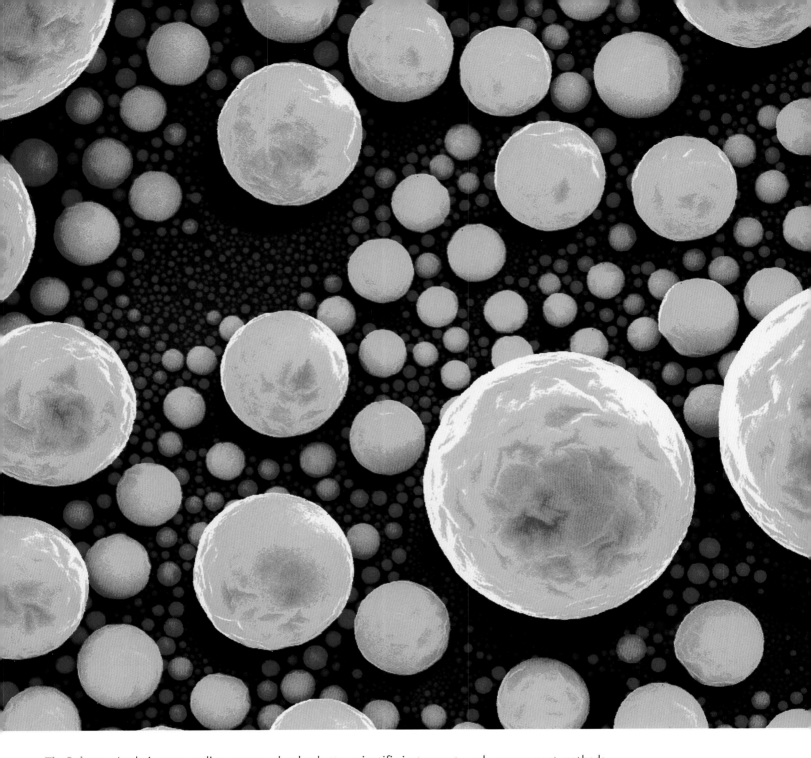

Tin Spheres In their never-ending quest to develop better scientific instruments and measurement methods, researchers at the National Institutes of Standards and Technology in Gaithersburg, Maryland, upgraded a scanning electron microscope so that it would reveal hundreds of times more distinct light-intensity levels than could previous instruments. This instrument produces images of microscopic samples that are strikingly vivid and harbor a stunning sense of depth. In this image, the instrument captured molten tin particles that had solidified into spheres of various sizes; the colors, which correspond to the different so-called gray-level intensities of the raw image, were added later. By studying the distribution of sizes and surface features of these metallic spheres, the researchers were able to assess the imaging performance of their instrument. The largest spheres are about 1.5 microns (1.5×10^{-6}) in diameter; the smallest span about 100 nanometers (10^{-7} meter).

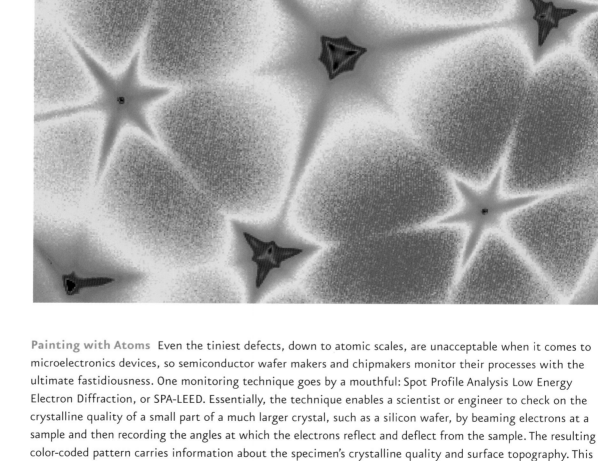

Painting with Atoms Even the tiniest defects, down to atomic scales, are unacceptable when it comes to microelectronics devices, so semiconductor wafer makers and chipmakers monitor their processes with the ultimate fastidiousness. One monitoring technique goes by a mouthful: Spot Profile Analysis Low Energy Electron Diffraction, or SPA-LEED. Essentially, the technique enables a scientist or engineer to check on the crystalline quality of a small part of a much larger crystal, such as a silicon wafer, by beaming electrons at a sample and then recording the angles at which the electrons reflect and deflect from the sample. The resulting color-coded pattern carries information about the specimen's crystalline quality and surface topography. This SPA-LEED image was generated during a process in which new silicon atoms were being deposited atop a silicon wafer. From the data, scientists could discern that the originally smooth surface had reformed into a hill and valley microscape with pyramidal hills averaging about 2 nanometers in height and 20 nanometers in width. The data harbor information on structures ranging in size from 3 angstroms (3×10^{-10} meter) to 100 nanometers (10^{-7} meter).

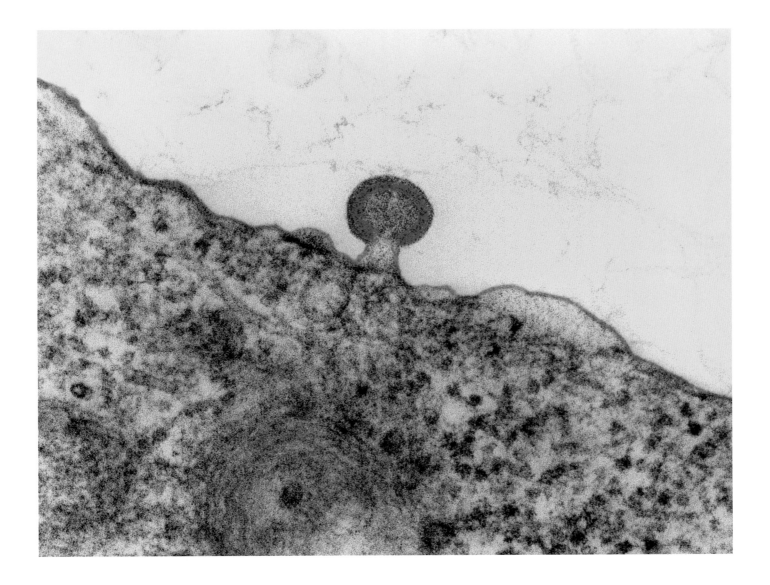

Viral Sunrise Like a sun over turbulent waters, a single human immunodeficiency virus (HIV)—the one that causes AIDS—buds directly from the membrane of a cell in a sample of human lymph tissue. In this image, a microscopist used a transmission electron microscope, which generates images using electrons passing through samples, to reveal the virus in the act of emergence. An extension of cell membrane (shown as a light-blue border) is in the process of pinching off to release the virus. The blue area coincides with the cell's cytoplasm, and the yellow demarcates the exterior of the cell. The HIV particle has a diameter of about 100 nanometers (10^{-7} meter).

Self-Cleaning Pipe Equipped with tiny cilia, or mobile hairs, the lining of the trachea, or windpipe, functions like an automatic sweeper. Working together, the hairs clear away fluids or particles. The larger hairs are a few microns long, which means they might span the length of a bacterium, and they're less than one micron in diameter, which makes them roughly 1/70 as thick as a typical head hair. The smaller cilia, or microvilli, visible in this scanning electron micrograph are about 1 micron long and about 100 nanometers in diameter, which makes them just a bit thicker than a cold virus (10^{-7} meter).

Silicon Phyllo A major quest in the science of materials has been to reveal the structures of things on all relevant architectural levels, ranging from the atomic and molecular constituents of materials up to the entire material, whether it be a silicon wafer or a steel I-beam. Here, materials scientists have prepared a silicon crystal so that portions of its atomic planes are unencumbered by overlying ones, resulting in a step-like structure. Then, using a scanning tunneling microscope, the researchers were able to harvest an image of these silicon steps down to the atomic level of detail. The ultimate pointillism of materials becomes beautifully apparent. Darker spots correspond to tiny areas of the crystal that harbor either defects or atoms other than silicon. Individual dots correspond to pairs of silicon atoms. Each step is about 1.4 angstroms high (1.4×10^{-10} meter) and the entire stretch visible here spans just under one-fifth of a micron (2×10^{-7} meter).

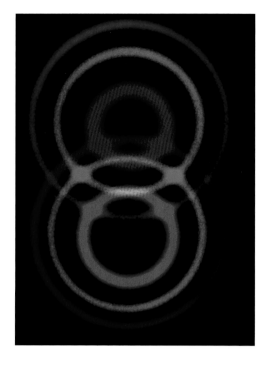

Entangling Light becomes ever more fantastic the more we learn about it. Quantum mechanicians have found that they can "entangle" two photons, the particles that make up light. In other words, even though the photons are separate, they can share a single quantum state, which is like saying that one photon carries the face of a coin and the other photon carries the tail. Actually, it's weirder than that because both photons are, in a sense, simultaneously in the head state and the tail state until someone takes a look at one of the photons. That act of looking effectively causes the photon to "choose" one or the other state, and that instantly entails that the other photon carries the partner state. Thus when scientists make a measurement on one entangled photon, they instantly know the state of the other even if it is a billion miles away and out of reach of any detector. These sorts of situations used to be relegated to the realm of thought experiments, but researchers recently have been producing real entangled particles in laboratories. This composition of luminescent colored circles was produced in an experiment in which researchers sent the beam of an ultraviolet laser through a special crystal that converts the beam's relatively shorter wavelength photons into pairs of lower-energy daughter photons with longer wavelengths. This one-to-two split shows up as the two sets of colored concentric rings. Under some conditions, one of the daughter photons will be polarized horizontally (its electromagnetic waves oscillate in a horizontal plane) and the other will be polarized vertically (its electromagnetic waves oscillate in a vertical plane). If these are made to intersect, as they do here, the daughter photons become entangled—the photons have no definite polarization, yet each member of an entangled pair, when measured, still will harbor one or the other type of polarization. Cryptographers are hoping to exploit entangled photons like these in new secret messaging systems. Ultraviolet photons have wavelengths in the 200-nanometer range ($\sim 2 \times 10^{-7}$ meter).

Atomscope Since the early 1990s, scientists have been developing a family of amazing detail-oriented tools known as scanning probe microscopes. Most are based on an exquisitely fine stylus that is made to sweep back and forth over a surface. That action generates a continuously changing signal—such as a tiny electrical current or a mechanical deflection in the stylus—that correlates with a physical or chemical feature of the sample underneath the probe. A computer keeps track of all the changes and then reconstructs an image of the underlying sample, much as a blind man might reconstruct in his mind an image of a face by touching all parts of it. Scanning probe microscopes have enabled researchers to visualize the atomic and molecular landscapes of the world as never before. It took a different kind of tool, a scanning electron microscope, to visualize the stylus of one of the newer members of the scanning probe microscopy family, the scanning gate spectroscope. This probe relies on one tip to generate topographic maps of surfaces with nearly atomic resolution, while its other tip simultaneously serves as an electronic gate with which researchers can measure how electricity behaves in tiny regions of the underlying surface. This kind of dual information could help engineers learn how to fabricate ever-more-miniaturized circuitry, among other things. The color was added to the image to highlight the stylus' features, including the metal electrodes (in gold) for both controlling the probe and gathering its readings. The two tips of the probe are separated by a gap of about 200 nanometers, about three times the width of a typical cold virus (2×10^{-7} meter).

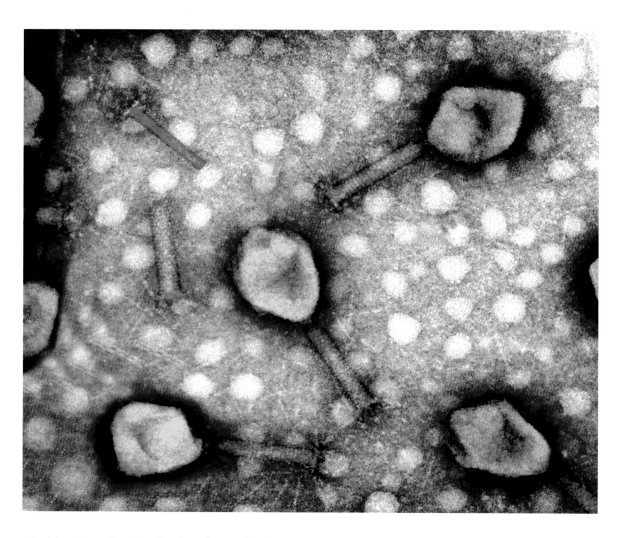

Machina Genetica Smaller than the smallest bacterium, viruses, such as these T4 bacteriophages, are among nature's tiniest packages of genetic material. Their yellow heads—twenty-sided structures made of protein facets—carry strands of DNA that encode genes for making more T4 bacteriophages. The natural machinery of these entities is apparent in this transmission electron micrograph. First the green-pronged anchor at one end of the bacteriophage lodges into a bacterial cell, and then the blue shaft between the anchor and the yellow head undergoes a remarkable structural transformation related to the one that renders certain iron alloys into steel. In a kind of twisting action, the protein molecules that make up the shaft shift their relative position, thereby suddenly shortening the shaft. Like a plunger of a syringe being pressed, this process forces the DNA from the head through the shaft and into the underlying bacterial cell. Once inside, the DNA ensconces itself in the host's own genetic machinery, which then expresses the T4's genes. The end result: more bacteriophages. A microscopist sent a beam of electrons through a preparation of T4 bacteriophages to obtain this image and added the colors later to highlight the viruses' different parts. Each bacteriophage is about 200 nanometers long (2×10^{-7} meter).

Self-Assembly A popular sport among molecular scientists these days is to develop techniques in which molecules will spontaneously arrange into ever-larger structures that are scientifically or technologically useful. After all, it is by this sort of self-assembly on the molecular scale that biological structures form. This bottom-up approach to making things also could prove useful for making molecule-scale electronic components that are far smaller than the ones populating today's most capable microprocessors. In this image, researchers used an atomic force microscope (AFM) to visualize the kinds of structures that self-assemble when arene molecules—a class of organic molecules that typically have a ring-like structure with one or more small or linear chemical appendages—are placed onto a surface of crystalline graphite. AFM relies on a tiny stylus that scans just above a sample's surface. As it does, the stylus responds to the tiny pulls and tugs exerted by the underlying sample's chemical composition or topography at different points and a computer then reconstructs the underlying surface from those responses. This particular preparation of arene molecules had a tendency to bifurcate during the self-assembly process. Since the branches of this process can rejoin other branches, the overall structure ends up as a loose mesh through which some of the graphite surface can be seen in purple. Each side of this scan is about 10 microns long (10^{-5} meter); the widths of the arene branches are in the range of 100s of nanometers ($\sim 10^{-7}$ meter).

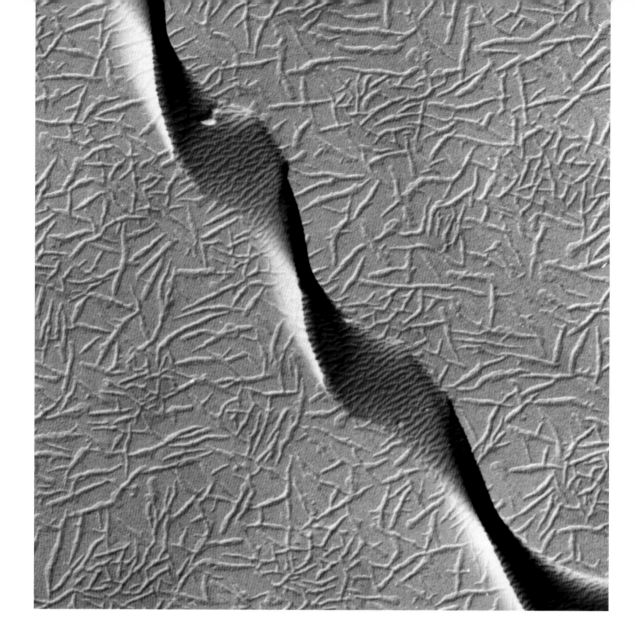

Twisted Crystal The polystyrene molecules in your disposable cup, the actin protein molecules in your bicep, and the DNA in your cell nuclei are all polymers. These sometimes enormous molecules are made of smaller building blocks, or monomers, that bond together in linear, branched, or networked structures. Make a polymer out of amino acids and you end up with proteins like actin. Make it out of synthetic organic monomers, such as styrene or ethylene, and you can end up with familiar plastics. When a group of polymer scientists at the University of Akron, which is smack in the middle of the United States' "Polymer Valley," incorporated a key feature of natural monomers into their synthetic polymers, they ended up with tantalizing helical shapes reminiscent of biological polymers. This gorgeous helix, which is about one-hundredth the width of a human hair, is made of thousands upon thousands of layers, or lamellae, like an enormous stack of cards. Each lamella consists of about two thousand monomers lined up side to side with the molecules' long chains swerving to form many S-curves. The overall twist is the result of each successive lamella stacking in a slightly rotated position with respect to the adjacent ones. Moreover, the molecules within each lamella are themselves systematically rotated with respect to one another, leading to a significant warp in each layer. The structure is completely novel, and novel structures always hold promise for such payoffs as new materials with unprecedented capabilities. The helix is about 500 nanometers wide (5×10^{-7} meter).

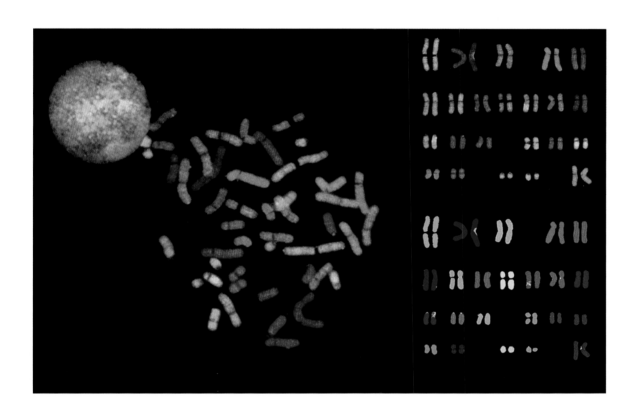

Painting Chromosomes Each chromosome carries a characteristic roster of genes. By making small genetic snippets that bond to specific genes, and by attaching to those snippets molecules that fluoresce in different colors, it has become possible to essentially paint each of a human being's twenty-three distinct pairs of chromosomes a different color. When the colored snippets, or probes, are bathed over a cell during its interphase—when the cell is replicating and its DNA is bunched together—the result is a multicolored cell nucleus. But as the cell divides and its chromosomes replicate, the differently colored chromosomes can be distinguished in a microscope, electronically photographed, and then digitally arranged, almost as in a police lineup, so that each chromosome can be examined. This technique, known as spectral karyotyping, or SKY™, can be useful for spotting chromosome abnormalities associated with diseases. For example, if a translocation occurs—when a piece of one chromosome becomes part of another—it might show up in a SKY analysis as a little band of, say, green, on an otherwise red-labeled chromosome. To get the image above, researchers used a custom-designed set of optical filters that enabled a light detector known as a spectrophotometer and an imaging chip, much like the ones in video cameras, to selectively detect the different wavelengths from the differently colored chromosomes. Each arm of a chromosome in this image is about 700 nanometers in diameter (7×10^{-7} meter).

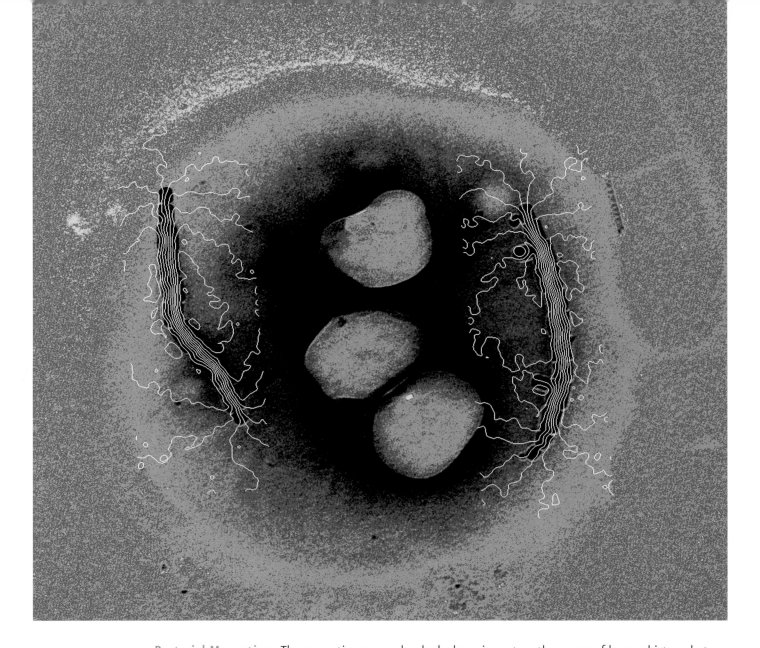

Bacterial Magnetism The magnetic compass has had a huge impact on the course of human history, but biology's simpler creatures have been applying the principle behind the compass for billions of years. Just like iron filings near a magnet, magnetotactic bacteria placed near a magnet will spontaneously arrange in linear patterns that reveal the magnet's lines of force. That's because these bacteria manufacture magnetosomes, tiny crystals of magnetic minerals. Aligned like pearls on a necklace, these magnetic particles enable the microbes, such as this lagoon-dwelling specimen, to orient themselves in accordance with the Earth's magnetic field. If water currents or other nearby activity dislodge such bacteria from, say, their normal nutrient-rich locations at the bottom into less-friendly waters above them, the microbes use their magnetic sensibility to navigate back to the nutrients. In this image of a magnetotactic bacterium that was found in a small lagoon at Sweet Springs Nature Reserve on Morro Bay in California, the magnetosomes appear as black and gray stripes on either side of the bacterial cell. Most remarkable in this image is that the magnetic field lines are rendered visible, as squiggly white lines. This unusual sight is made possible with an innovative technique referred to as off-axis electron holography, which is a specialized form of electron microscopy. This image, by Rafal Dunin-Borkowski, of the University of Cambridge, was a winner of the Novartis 2001 Visions of Science Competition. The bacterium is one micron or so in diameter (10^{-6} meter).

Reading Metal Some metals are better than others at being pressed, bent, and formed into shapes. Brass is one of them, which is why, for example, it's a metal of choice for making shells for munitions. The formability of brass, an alloy of mainly copper and zinc, comes down to its microstructure—the way its crystal grains behave both individually and with respect to each another. In short, for a metal to take on new shapes without tearing or failing in some way, its grains have to change their shapes and orientations. And for that to happen, the atomic planes of the crystals themselves must undergo transformations. These details often show up in a 150-year-old technique known as metallography, in which a metal specimen is sliced to have a smooth surface and then etched with acid to bring out the structural details of the grains at the surface. Here, a number of irregularly shaped grains are packed together. Visible in this image are scores of dark parallel slip lines, which are layers of metal that slip along one another, like sheets of paper. The lines stop dead at the boundaries of the crystal grains that make up the metal. Microstructural details like these enable metallurgists to track the quality and behavior of alloys. The distances between the slip lines probably are on the order of microns ($\sim 10^{-6}$ meter).

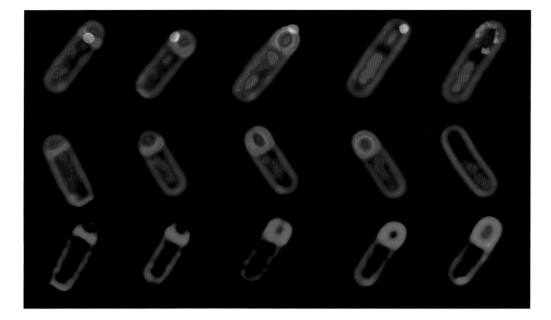

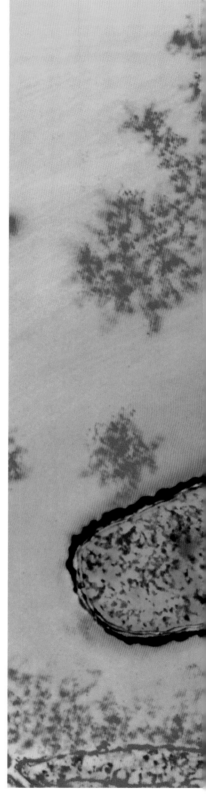

Microbial Self-Preservation When some bacteria find themselves near starvation, they initiate a cascade of events that leads to their transformation into spores that can last for years until the conditions become conducive to a good life. Spore formation, or sporulation, begins when the bacterium replicates to create a two-cell structure. The new cell is called a forespore and is separated from the parent cell by a thin septum. The edges of this septum then migrate upward all around the forespore, ultimately pinching off completely to create a cell totally encapsulated within the mother cell. This process, known as engulfment, requires a lot of molecular orchestration and action, including the transportation of proteins and other big molecules from one place to another. To help unveil what types of proteins, or molecular machinery, is involved in the engulfment process, scientists have used fluorescent dyes to label specific cellular components. The labeled microparts then can be readily seen with a fluorescence micro-scope. In this image *Bacillus subtilis* cells are undergoing sporulation. In the top two rows, the blue marks the location of chromosomes, and membranes were labeled with a fluorescent red dye. In the top row, the bright green spot in each sporulating bacterium flags the presence of a protein that is central to the forespore engulfment process. The middle row and bottom row (with only a green stain) show more examples of engulfment as the process unfolds from left to right. Each bacterium is about two microns long (2×10^{-6} meter).

Bioterrorist In the days after September 11, 2001, a second wave of terror swept the country: Letters loaded with deadly anthrax spores, which were modified and processed to increase their ability to infect more people, were handled by the U.S. postal system. The anthrax spread in both predictable and puzzling ways. What followed was sickness and death on a relatively small scale, but fear on a national scale. Shown in this falsely colored scanning electron micrograph is a single *Bacillus anthracis* cell. Anthrax normally infects and kills livestock, such as cows and sheep, by causing pneumonia. In the past, people have been infected by anthrax only after handling the hair, hides, and other tissues or remains of animals that had been stricken. Germ-weapons programs in the United States, the former Soviet Union, Iraq, and perhaps elsewhere changed that and turned anthrax into a biological weapon, and terrorists capitalized on that example. This anthrax cell is about two microns long (2×10^{-6} meter).

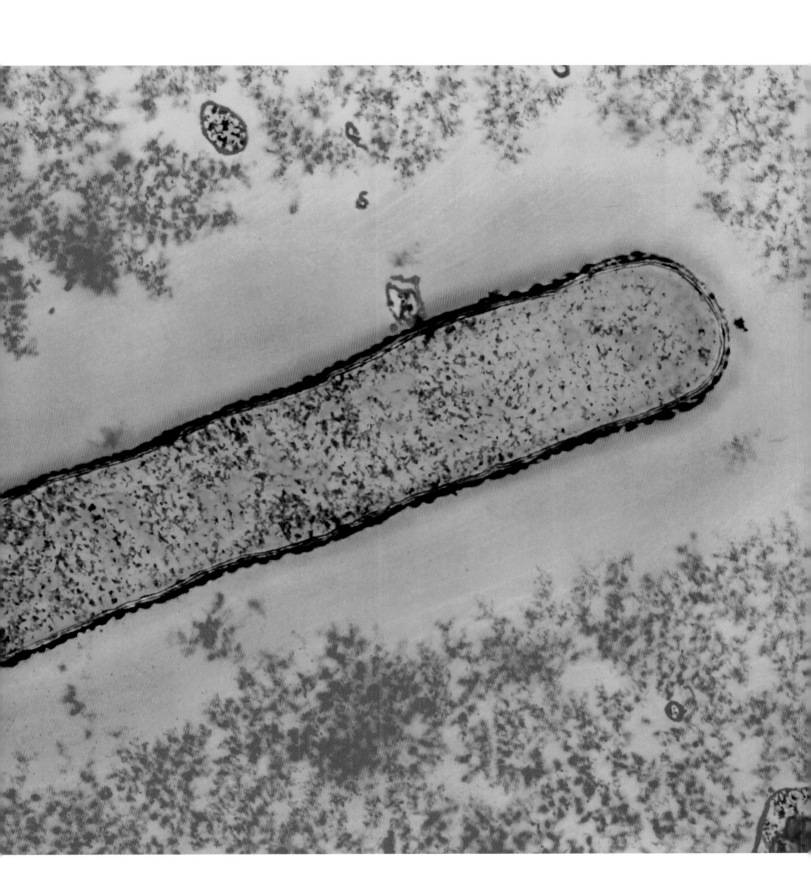

Arachnida's Silk Road To weave its web, the spiny-backed spider brews up a concoction of liquid protein in its silk glands. As the liquid is pushed through the glands' cylindrical spigots located on the posterior abdomen of the spider (shown as striated red-orange structures with a tree-trunk-like base and hose-like ending), the constituent protein molecules are forced to align as they solidify into tough, stretchy fibers (here, colored in green and blue) on their way out of the spigots. These then are woven into the silk threads the spider uses for webs, for encasing victims, and for climbing. By mixing and matching different protein ingredients before spinning their fibers, spiders can make a variety of different silks, each with properties matched for its particular purpose. Each of the strands here is about 3–5 microns in diameter (3–5 x 10^{-6} meter).

Cardiac Plaid Muscles are among the most fantastic of molecular machines. The basic unit, known as a sarcomere, consists of a pair of comb-like structures—made largely of filaments of the proteins actin and myosin—that face each other and whose prongs closely interdigitate. When muscles flex, a ratcheting mechanism in each sarcomere causes one set of prongs to slide along its partner set, causing the overall sarcomere structure to get thinner. Arranged in a multitude of columns and sheets, the sarcomeres together make muscles move, in everything from pumping hearts and masticating tongues to sprinting legs. In this transmission electron micrograph, an array of sarcomeres within a single heart cell is visible in the yellow, orange, and red stripes. The gray-blue structure in the middle is the cell's nucleus, and the red splotches are mitochondria, the cellular organelles that refine food-derived molecules into adenosine triphosphate (ATP), the chemical fuel that drives many cellular functions including muscle action. The banding of the sarcomeres directly reflects the structural alignments of the proteins that make up these structures. The length of each sarcomere, which spans between pairs of thin red lines, is about 2.5 microns (2.5×10^{-6} meter).

Microstove With some of the same techniques used for making microelectronics, researchers have been fashioning a menagerie of microelectromechanical systems, or MEMS. Little MEMS-based motion and pressure sensors are already in such venues as car air-bag triggering systems and blood-pressure monitors. Many MEMS aficionados envision their work someday amounting to a new machine age writ small. Like big machines, MEMS have tiny parts that vibrate, flex, rotate, light up, and, in this case, heat up. The back-and-forth structure colored in orange is a heating element that's about the size of a bacterium. Inject different amounts of current into this tiny system, and it heats up to a greater or lesser degree. MEMS arrays like this one could be used to calibrate such gadgetry as infrared cameras and military sensors in the nose cones of missiles. The heating element is about 4 microns across (4×10^{-6} meter).

Hair that Hears Tiny hairs, exquisitely fine ones arranged in beautiful functional structures, are the basis of most of the hearing that goes on in the animal kingdom. This stoic little bundle of hairs, the tallest only a few microns high, bends and sways like grass in response to vibrations. This particular bundle is rooted in a single sensory cell from the inside of the ear of a turtle. As it moves back and forth, its mechanical motion alternately stimulates the sensory cell and inhibits it from sparking a nervous impulse toward the brain's auditory processing centers. The image was produced with a scanning electron microscope. The bundle reaches up just a few microns ($\sim 10^{-6}$ meter).

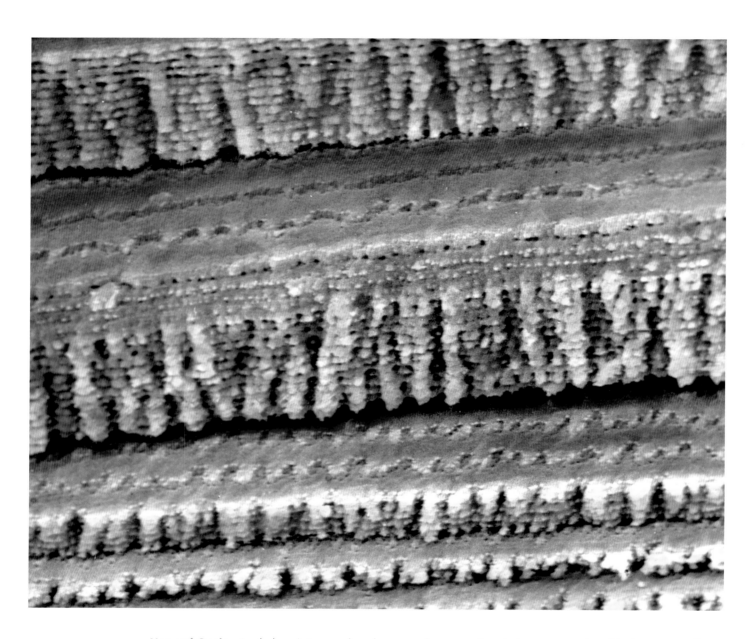

Natural Optics Look deep into your lover's eyes and you see directly into a mysterious blackness, actually a series of solid and liquid tissues. One of those is the lens, which is one of the most remarkable natural examples of how function follows form. About as thick as a nickel, it is made of cells devoid of nuclei and with cross-sections resembling squashed hexagons. The cells are packed with a family of proteins known as crystallins, which, about two thousand times longer than they are wide, are fiber-like. In this falsely colored scanning electron microscope image, each of the discernible vertical layers corresponds to a portion of a single lens cell. Each long, flat lens cell is 5 microns thick (5×10^{-6} meter).

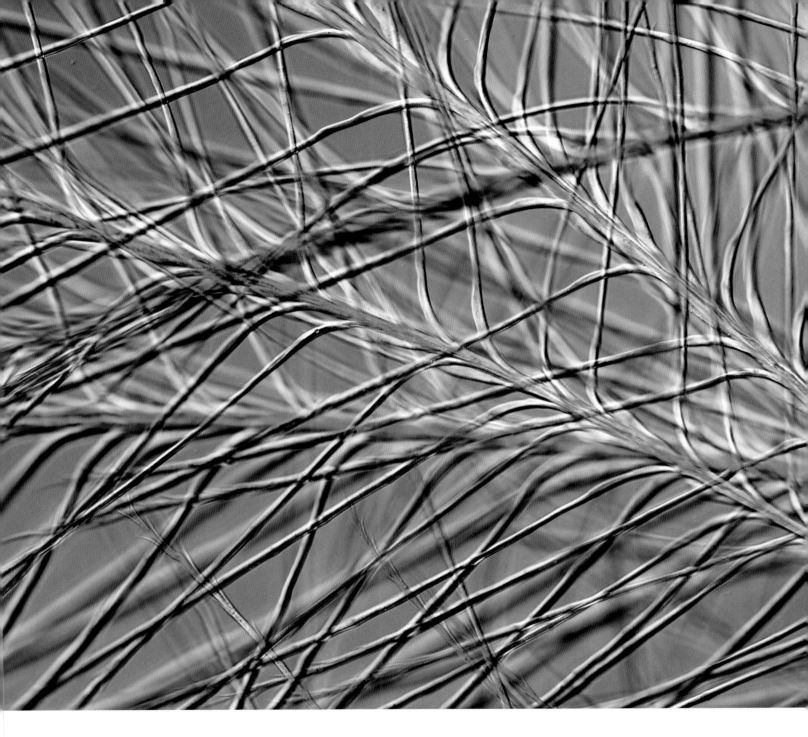

Deep in Down Feathers cannot be beat when it comes to nature's knack for marrying form and function. Specialized epidermal growths, they're made largely of the protein keratin fashioned into lightweight and flexible structures that can withstand the rigors of wind and flight. Down feathers, such as the ones in this photomicrograph, usually serve to insulate as they lie underneath the contour feathers on the wings and backs of birds. Visible here are the main stems—or quills—of several down feathers and the many thin barbs that extend from them. On contour feathers, the barbs end in barbules, which help adjacent barbs to hook together into stiff, regimented structures. Down feathers do not have such barbules, so their barbs sum into the fluff that makes down so soft and suitable for pillows. Add its insulation powers to its fluff, and down's popularity for winter wear also becomes apparent. The smallest fibers here are about 5 microns thick (5×10^{-6} meter).

Steel Abstraction Take a piece of steel, slice it, and polish its surface to a nice shine. Then etch it with an acid that eats downward at different rates depending on the local compositional and physical differences in the metal. Now look at it under a microscope, as a metallurgist has done in this case. The result is a framable composition of forms and patterns, one that also tells metallurgists about the quality of the metal. In this case stainless steel—made predominantly of two types of metallic crystals—shows up as a lamellar structure. The distances between the layers range from 5 to 10 microns (5×10^{-6} meter–10^{-5} meter).

Electronic Grace The locations and types of atoms in a layered semiconductor crystal influence the electrons'
moment-to-moment trajectories the way attractive and repulsive people or storefronts influence the pathway
of a city walker. Under some conditions, those electrons form into a veritable sheet of gas that spreads
through one of the layers. If those layers are "bumpy," which means they might contain kinds of atoms whose
sizes and charges will deflect electrons moving near them, the resulting electronic pathways can take on
stunning forms, at least when rendered the way Harvard physicist and artist Eric Heller has chosen to. In
this simulation, and in real semiconductors, the bumps shunt the electrons along specific paths, or branches.
In this rendering, electrons enter the semiconductor from the center and then radiate out, spontaneously
forming branches. To Heller, the beautiful pattern is reminiscent of "translucent kelp" and "the ridges on a
mountain." This quantum-scale phenomenon was not discernible in the experimental data; it became apparent
only through the visualization that is possible in computational studies. Understanding this channeling or
branching could prove to be a benefit to engineers designing the future's ever-smaller electronic devices.
Each side of this simulated semiconductor is about 5.5 microns long (5.5×10^{-6} meter).

Wires sans Resistance The electrical conductivity of materials ranges from little or none—in which case the material is an insulator—to superconducting substances that carry electricity with absolutely no resistance under certain extreme conditions. Shown in cross-section here—at two different degrees of magnification—is a prototype superconducting wire made of thousands of filaments of a niobium-titanium alloy, which becomes a superconductor when chilled to or below −269° C, embedded in a matrix of copper metal. These kinds of wires are key components in sensitive detectors of magnetic fields and in the generation of the strong magnetic fields required for research in high-energy physics and fusion, and for such gadgetry as the magnetic resonance imaging machines that have become a standard tool for medical diagnoses. Each filament of the superconducting wire is 6 microns in diameter (6×10^{-6} meter); the wire's width comes to about .65 millimeters (6.5×10^{-4} meter).

Chalk Reincarnated Calcium carbonate, or calcite, can crystallize into many beautiful forms. Here is one —a hierarchy of cubes. The entire crystal structure, visible through a scanning electron microscope, is equivalent in length to about ten wavelengths of red or red-orange light. The crystal's volume is roughly equivalent to that of a small biological cell. Each side of the cube is about 6 microns long (6×10^{-6} meter).

Cells that See A retina is an astounding sensor whose complex, multilayered structure harvests optical signals from the world and begins the neural processing by which those signals become meaningful to the retina's owner. To visualize some of the retina's cellular layers, researchers developed antibody molecules that specifically bind to the double-cone cells that take in incoming light and convert that energy into nervous impulses that travel to deeper parts of the visual system. With the help of a stain that links specifically to the antibodies, the overall structure and orientation of the double-cone cells (here, in the retina of a zebrafish) becomes visible. The two ends of the cells also have been adorned with a red molecular label. The combination of this label and the green-labeled antibody yields the bright yellow at the ends of the double-cone cells. Meanwhile, the red dye also stains the outer segments of rod cells, the retina's other major type of light receptor, and the retina's inner plexiform layer, where enormous numbers of synaptic connections form to preprocess signals that then move into the optic nerve and into the more central portions of the visual system. The nuclei of the rod cells are dyed blue, but are themselves hardly visible since the inner fibers of the double-cone cells appear blue as they traverse them. The image required a triple exposure, each one obtained with a filter matched to image the structures labeled with the different dyes. Cone cells are about 6 microns in diameter (6×10^{-6} meter).

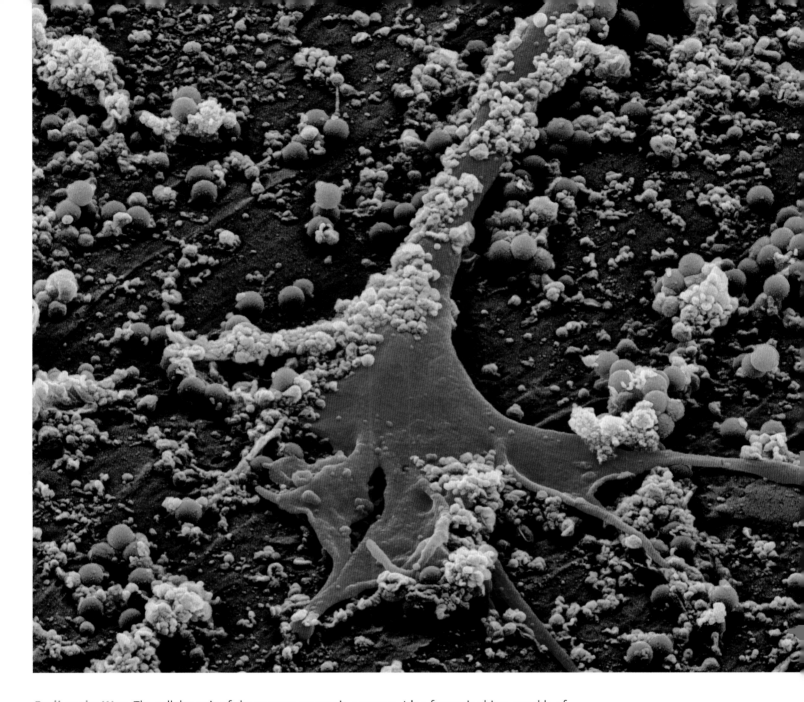

Feeling the Way The cellular unit of the nervous system is a neuron. It's a fantastic thing, capable of receiving electrochemical inputs, sometimes from many thousands of other cells, and then integrating it all into an output of its own. That outgoing signal carries a more streamlined and information-laden signal than any signal it received. When neurons first wire up with one another, they send out so-called growth cones at the ends of their long axons, the conduits that convey a signal from the cell's main body to its extremities where the cell interfaces with other neurons. With various molecular cues, those growth cones send out shoots that seek out the new connections. Here, a scanning electron microscope shows an isolated neuron's growth cone. The fluffy yellow stuff is membranous vesicles derived from the brain's optic tectum, which provides guiding signals to the migrating growth cones. The blue spheres are fluorescent silica beads of known size that provide landmarks so that researchers can measure and monitor how far the growth cones have migrated. The growth cone has a diameter of about 7 microns (7×10^{-6} meter).

Atomic Graffiti As technologists push their control over the material world into the nanometer scales, which is to say as they learn to manipulate the molecules of the world with the kind of finesse biological cells are famous for, they need to be able to visualize what is going on at those scales. Actually, the ability to see on nanoscales goes hand in hand with the ability to do things on nanoscales. A technique known as scanning Auger microscopy (SAM) is particularly suited for revealing very thin layers of one material on top of another. SAM works by shining X rays on the sample. Near-surface atoms absorb those X rays, become energized, and then vent much of that energy by ejecting electrons up and out toward a detector. The energy of these electrons will be altered by whatever atoms happen to be overlying them as they pass up and out of the sample. So by measuring the energy of those so-called Auger (pronounced oh-zhay) electrons, it is possible to infer what kinds of atoms interacted with them. That data then can be recast into an elemental map of the sample's surface and near surface. In this image, the blue corresponds to the underlying copper substrate on top of which had been deposited a thin and porous film of gold. Each spot, or pixel, in the image represents the composition of a virus-sized region of the surface. The area shown is 7.5 microns on a side (7.5 x 10^{-6} meter).

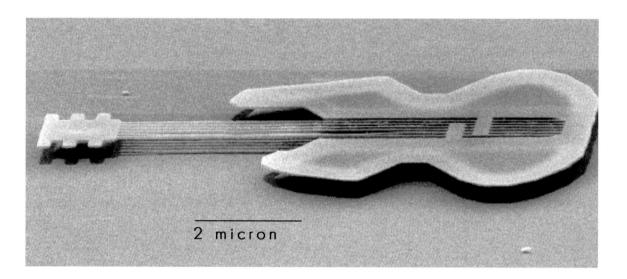

2 micron

Playing on the Nanoscale At only ten millionths of a meter in length, or roughly the diameter of a red blood cell, and with strings about 100 atoms wide, this microfabricated all-silicon structure has to be the smallest guitar in the world. If plucked, the strings would vibrate with an inaudible frequency of 10 megahertz or so, which corresponds to the electromagnetic frequencies of FM radio. Modeled after the Fender Stratocaster, the structure was made out of a silicon crystal using fabrication techniques adapted from the microelectronics industry. It is a particularly light-hearted illustration of a class of devices known as microelectromechanical systems, or MEMS, which have micromechanical as well as microelectronic components. Among MEMS that already were in use in the 1990s are accelerometers on chips that detect motions characteristics for crashes and chips with millions of tiny movable mirrors that are collectively able to project moving images onto screens. Many technology watchers expect MEMS to become as pervasive in the coming years as microelectronic devices have in the past few decades. The microguitar is about 10 microns long (10^{-5} meter).

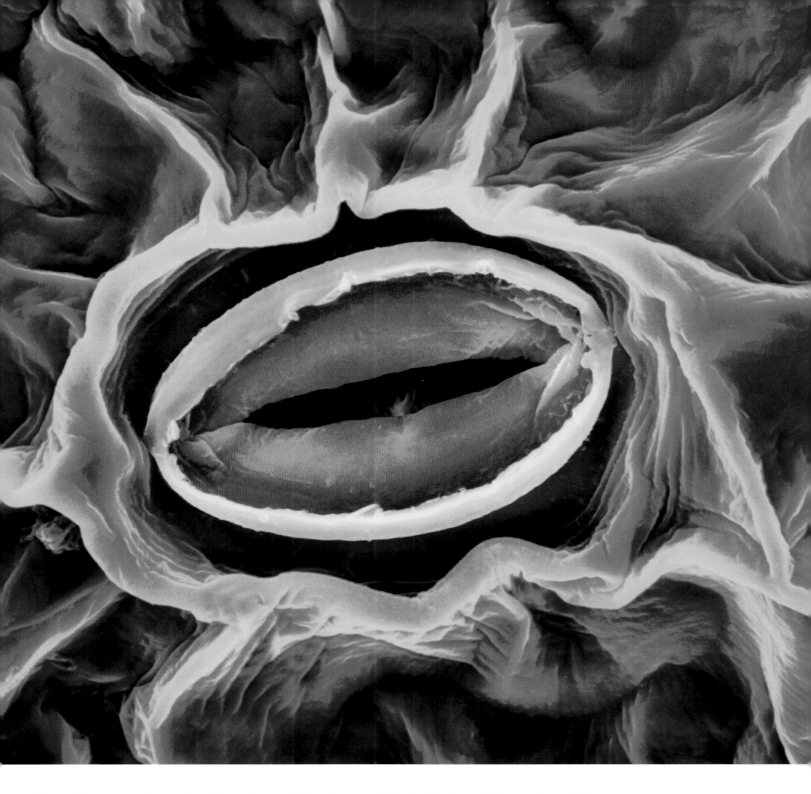

Secret Apertures Leaves, like this one from a dahlia plant, are riddled with tiny variable openings called stomates that open or close in response to the plant's moisture content and the humidity of the air. Adjusting the stomates is one of a plant's main means of controlling moisture and the exchange of incoming carbon dioxide gas and outgoing oxygen. Depending on the conditions, the cells on either side of the openings, called guard cells, highlighted here in orange, grow more or less turgid, thereby increasing or decreasing the stomate's openings. This image was taken with a scanning electron microscope. The stomate is about 10 microns across its small dimension (10^{-5} meter).

Flower Crystals Life is both soft and hard. Bones, teeth, and shells are familiar examples of biomineralization, a term referring to the many ways cells organize atoms and ions into hard stuff. Less familiar is the bio-mineralization of plants. Inside the cells of many flowering plants crystalline inclusions can be found; they come in many shapes, sizes, and chemical compositions, depending on the plant and its environment. Shown here is a type of inclusion known as a druse, a spherical mass of crystals of calcium oxalate, a chemical also used in ceramic glazes and found in human kidney stones. In the leaf cells of the *Peperomia astrid* plant, calcium oxalate druses can take on gorgeous, flower-like forms. The microscopist acquired this druse by slicing a leaf with a razor blade and dipping the cut region in a drop of water on a small piece of glass. After the water evaporated, the crystal was left behind for viewing in a scanning electron microscope. Druses may play many roles, including gathering light for photosynthesis. This one druse spans about 10.6 micrometers, which makes it roughly the size of a single red blood cell (1.06×10^{-5} meter).

Plastic Whirlpool Reminiscent of a satellite view of a massive hurricane, this is actually an atomic force microscope (AFM) image of a tiny circular crystal structure called a spherulite. Spherulites are created when some materials including polymers are melted and then allowed to solidify. This particular spherulite, about the size of a small biological cell, is made of the bacterially derived polymer known as poly(hydroxybutyrate), which, being biodegradable, could help alleviate the world's growing burden of plastic waste. The AFM relies on a tiny stylus that is deflected to a greater or lesser degree as it sweeps barely above a sample. By keeping track of the deflections, a computer is able to construct the image of the sample's surface, and color and shading are added to highlight a sense of depth. The spherulite's diameter spans about 12 microns (1.2×10^{-5} meter).

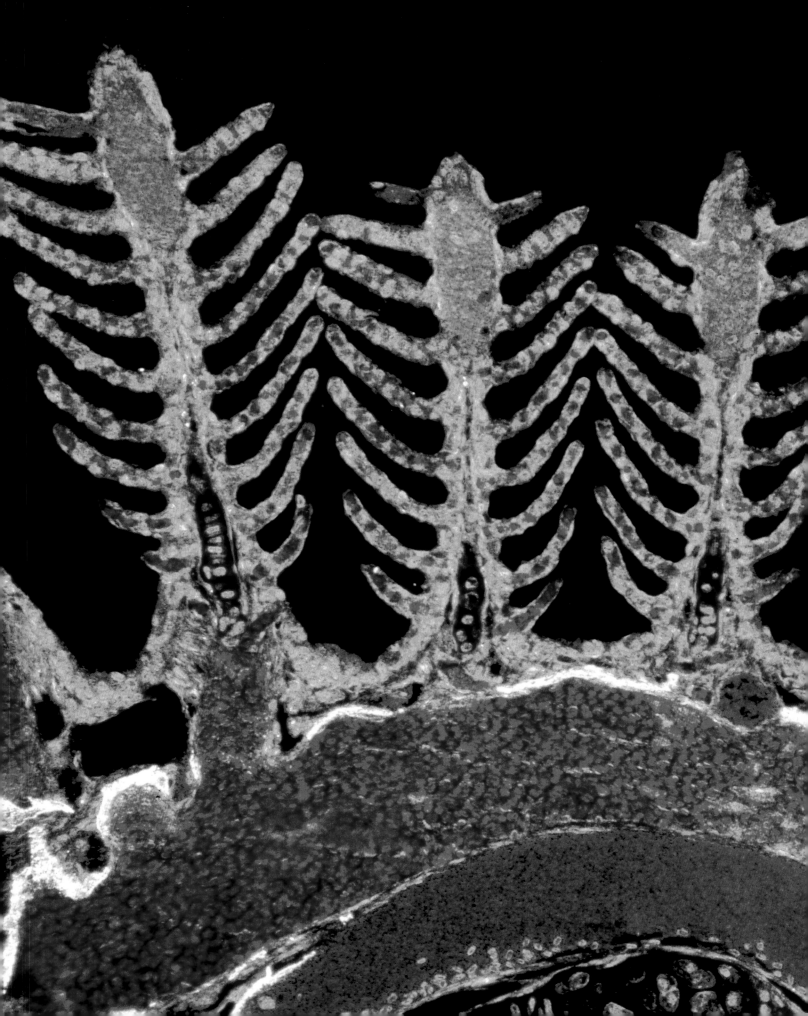

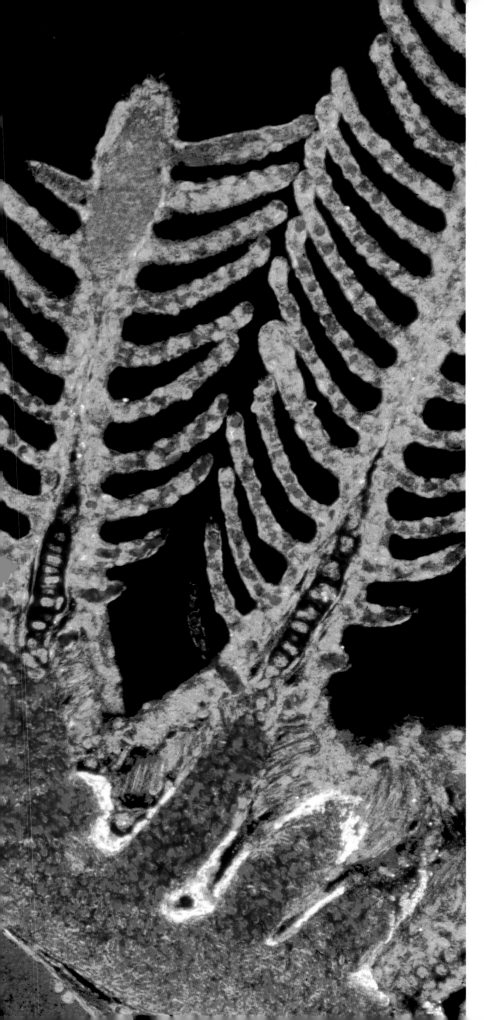

Natural SCUBA Breathing oxygen is a primary skill of the living and requires some sophisticated anatomy. One of biology's most widespread innovations for extracting oxygen from air (as well as dumping carbon dioxide and other waste products back into the air) is the lung. Its counterpart for water-dwelling creatures is the gill, which yanks oxygen out of water as the fluid rushes past the gill's enormous surface area. Looking like a stand of trees, these gill branches of a zebrafish (the fish version of a lab rat) transform the green deoxygenated blood into red, oxygenated blood. The image was obtained using darkfield illumination, which brings out color by allowing light to refract from the specimen's various structures and materials. Each "tree" is about 300 microns high, and the branches are about 12 microns wide, just enough for a red blood cell to squeeze through (1.2×10^{-5} meter).

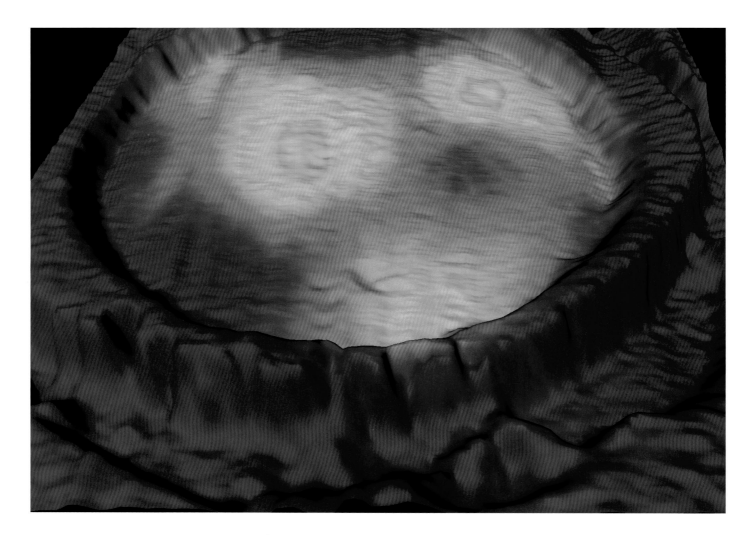

Mount Laserius In their quest to develop light-manipulating microtechnologies, or photonics, alongside microelectronic technology, researchers have been creating all manner of solid-state lasers, which are basically laser-beam–emitting crystals. These often are specialized multilayer crystals designed to shunt mobile positive and negative electric charges within the crystal toward one another so that these complementary charges meet with an emission of light. By flanking the light-emitting region of these crystals with additional crystalline layers that serve as mirrors for reflecting the light back into the light-emitting region, the entire construction becomes a laser. Here, researchers used an imaging technique known as near-field scanning optical microscopy to simultaneously image the topography of a vertical cavity surface emitting laser, or VCSEL, and the intensity of laser light emerging from different portions of the tiny structure. Yellow areas are the most intense. Reminiscent of a glowing crusty plug inside a volcano's crater, the colors and patterns here reveal how the VCSEL's detailed structure and its composition affects the light it generates and how that light makes its way out of the structure. Because VCSELs emit their light vertically, instead of out of the sides like most solid-state lasers, they hold promise for such uses as communicating between different layers of circuitry integrated into a single multilayer chip. The VCSEL is about 15 microns in diameter (1.5×10^{-5} meter).

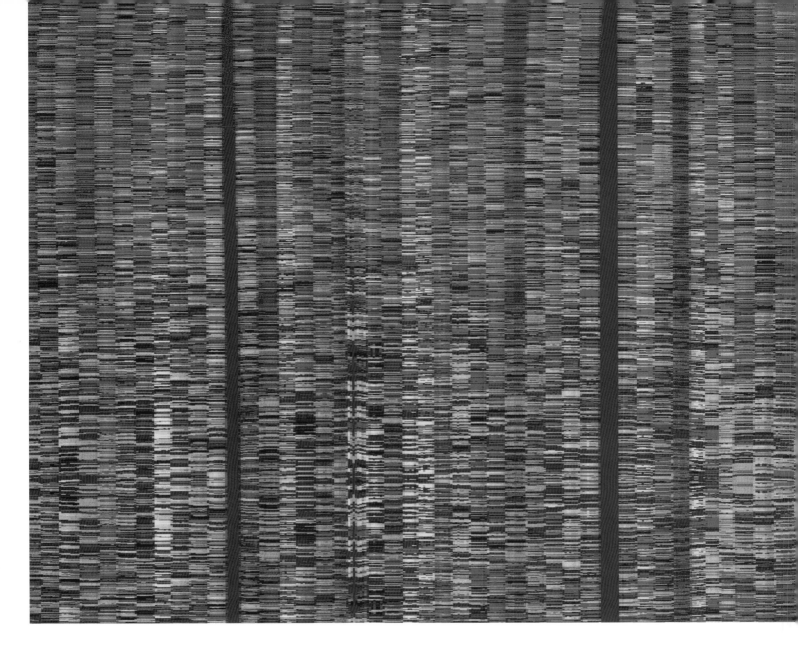

Reading the Code Throughout the 1990s, an army of researchers throughout the world worked at break-neck speed to do a first read on the raw genetic sequence of the human genome. The human genome, distributed over forty-six chromosomes, consists of tens of thousands of genes. Each gene is made of a long sequence of four types of genetic letters known as nucleotides, which the cell ultimately translates into the proteins that it needs to grow, reproduce, metabolize, and otherwise get on with the business of life. The genes themselves occupy only a small, still unknown fraction of the chromosomes, whose collective count of nucleotide pairs is roughly three billion. The image represents the kind of raw sequence data that genomics researchers have been generating. Each vertical lane reveals the sequence of a snippet of the entire human genome. Each of the little colored segments that make up each lane marks one of the four nucleotides. By labeling each nucleotide—known in letter form as A, C, T, and G—with a specific dye, it is possible to literally read off the genetic code as a long sequence from this four-letter alphabet. A computer analyzes the sequences from the lanes and looks for places of overlap, making it possible to reconstruct ever-longer portions of the genome. This image shows approximately 48,000 of the human genome's three billion nucleotides. If the length of DNA whose sequence is depicted were stretched out like a necklace, it would span nearly 16 microns (1.6×10^{-5} meter).

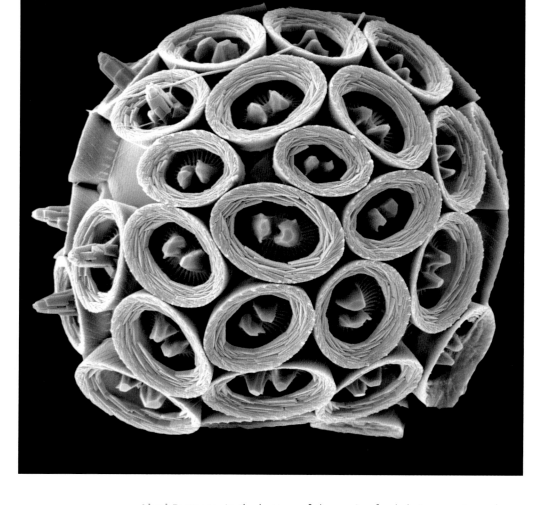

Algal Pottery At the bottom of the marine food chain are minuscule creatures known generally as plankton. Here micropaleontologist Markus Geisen photographed a single cell of the planktonic algae *Calcidiscus quadriperforatus* (right) undergoing one of the more dramatic makeovers that occurs in nature. This transition had never been photographed before. This type of algae predominantly reproduces asexually via cell division, but sometimes it switches to a sexual life cycle as well. When that happens, the resulting algae produces tough honeycomb-shaped calcite cloaks, known as holococcoliths, which are colored green in this image. When two of these cells fuse, the new generation of daughter cells produces the predominant form of coccolith, which looks somewhat like a lampshade with a central dimple. Shown here in brown, these structures seem to be held literally in the grasp of the honeycomb-like coccolith type. The image was taken with a field emission scanning electron microscope, which relies on a strong electric field that pulls electrons from the sample to a detector. The calcite artistry of plankton is as diverse as it is stunning. A scanning electron microscope close-up of the tiny marine species *Coronsphaera binodata* (above) provides but one more example. This organism is about 2 microns long (2×10^{-6} meter); the reproducing cluster of *Calcidiscus quadriperforatus* is about 17 microns in diameter (1.7×10^{-5} meter).

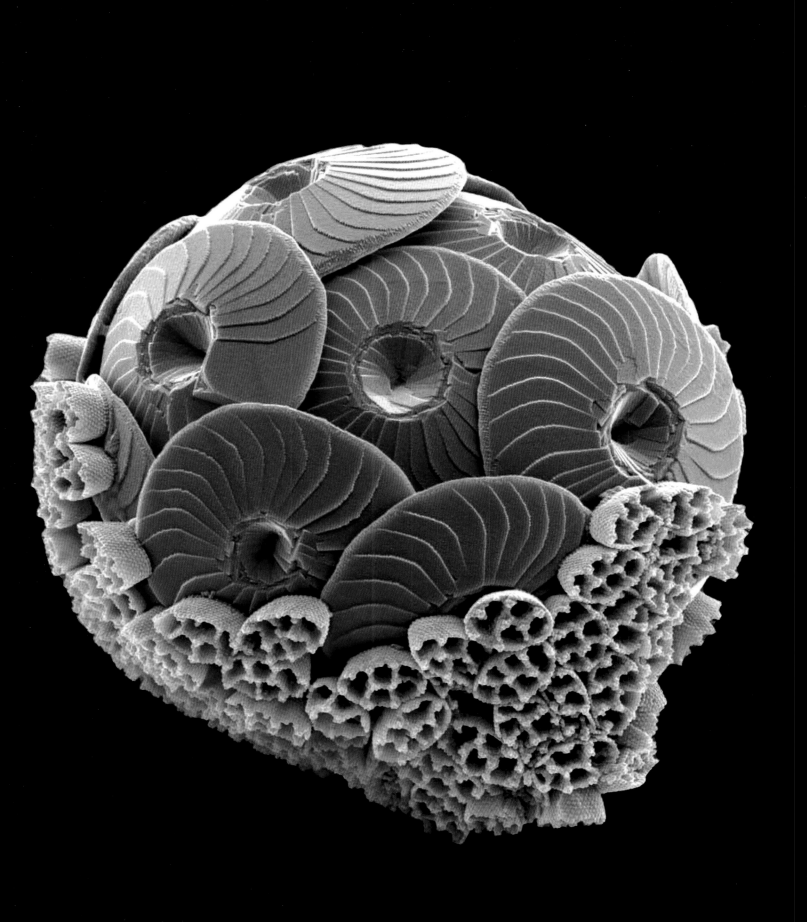

Romancing the Rose The rose petal, which is so smooth to the human touch, becomes a rugged microscape of moguls when viewed with an environmental scanning electron microscope, which unlike standard scanning electron microscopes, enables scientists to look at biological samples under realistic water-rich circumstances. When viewed with minuscule dew droplets that have nestled in and among the moguls, the grandeur of a rose petal endures on scales that are ordinarily invisible. The drops and dimples range in diameter from about 10 to 20 microns ($1\text{--}2 \times 10^{-5}$ meter).

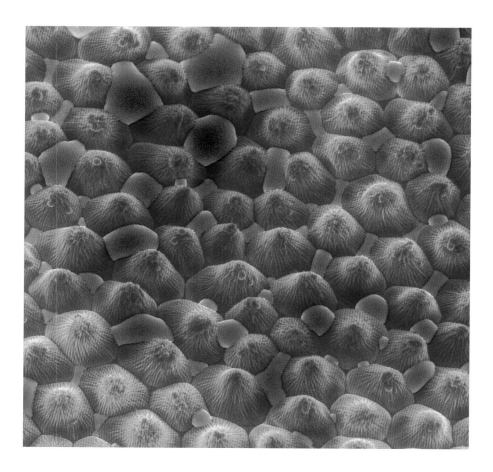

Planet Fat A biological cell is a tiny world enveloped in a membrane of fatty molecules. That membrane both separates a cell from, and links it to, its surroundings; its liquidity, permeability, and surface features influence what the cell takes in, what it exudes, and how it reacts to chemical and other signals coming from the outside. Cellular membranes are made largely of phospholipids, molecules with fatty appendages attached to hubs that point into the cell's watery interior on one side and to the watery exterior on the other side. The fatty appendages like to bunch together, and if there are enough of them they will spontaneously form into a spherical shape. Some researchers make models of cell membranes by mixing phospholipids together so they form structures called giant unilamellar phospholipid vesicles, or GUVs. This one is made of a mixture of cholesterol and two kinds of lipid molecules, one exactly four carbon atoms longer than the other. The longer lipids form an interconnected, filamentous, and semisolid network that cordons off larger, more liquid areas comprised largely of the smaller lipid molecules. By labeling the lipids with dyes that fluoresce with different colors when excited with specific wavelengths of light, biochemist Gerald Feigenson of Cornell University was able to produce a striking image that reveals the major distribution of a GUV's two lipid components. It was part of a study into the roles that cholesterol and cholesterol-related molecules play in cell membranes. The upper images were made with filters that highlighted the emissions from the short and long lipids respectively. The main image was created by merging the data from those images and then assigning the dye colors to the solid and the liquid regions. The GUV is about 20 microns in diameter (2×10^{-5} meter).

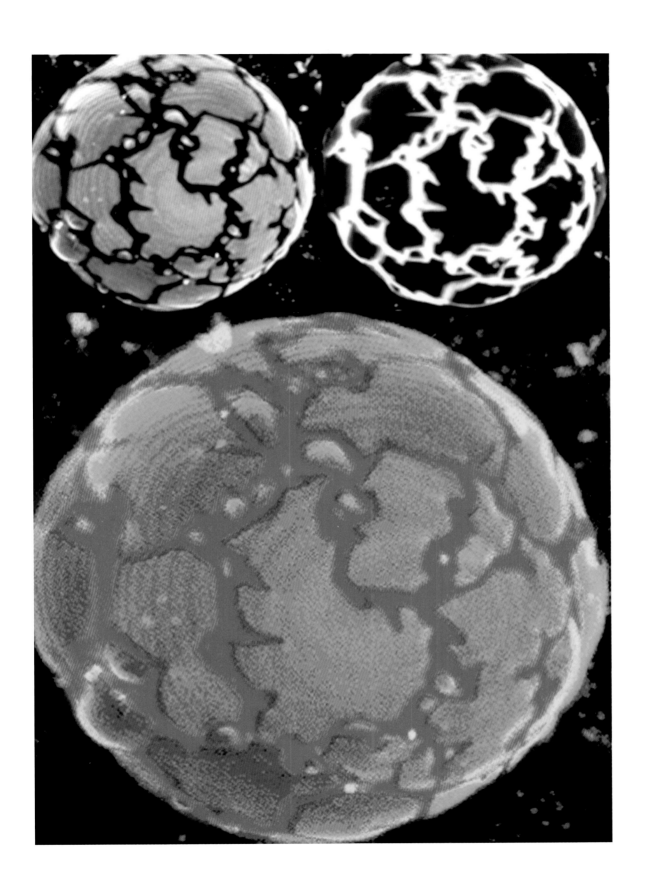

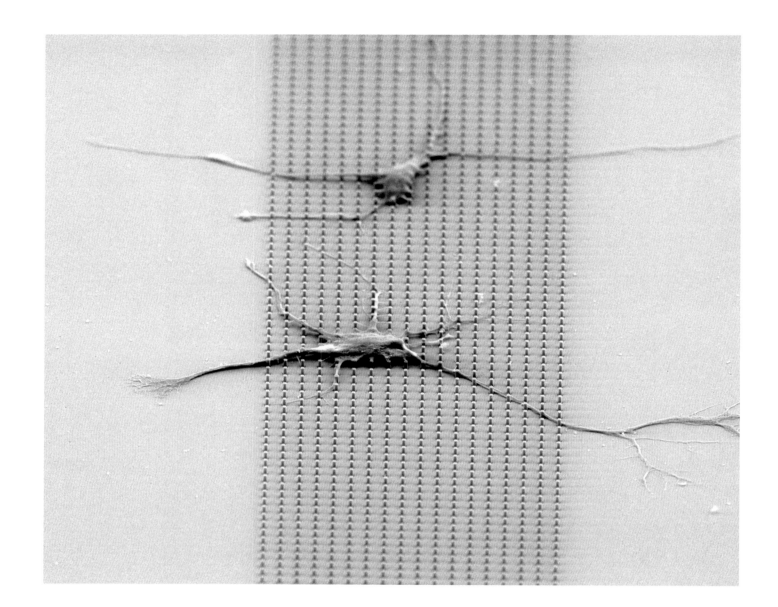

Brain Boulevard During the development of the nervous system, new brain cells, or neurons, send out extensions that ultimately link into a network capable of supporting the most miraculous phenomenon we know—the mind. Neurons self-organize with help from physical and chemical cues, which are themselves ultimately orchestrated by well-timed genetic and biochemical processes. To investigate the influence of physical surroundings on the growth of neurons, researchers placed neurons onto a microfabricated array of tiny silicon pillars, each about 750 nanometers (7.5×10^{-7} meter) in diameter. As it turned out, the outgrowths of the cells did conform to the geometry of the pillars, taking straight or diagonal pathways through the grid. By replacing the pillars with tiny electrodes, it is possible to study the way electrochemical signals are received by, travel through, and are processed by the anatomy of a single neuron. That kind of information can help neuroscientists better understand how brains work. The cell's bodies are about 20 microns in diameter (2×10^{-5} meter).

Fractured Clues Fractography—the science of breaking apart various specimens of metal and other materials to learn how their internal structures relate to the components of the materials, how those ingredients were processed, and the properties and capabilities of the resulting materials—is a powerful scientific and engineering analysis tool. Here, scientists used an atomic-force microscope to capture an especially detailed view of the fracture surface of a broken piece of epoxy resin. This type of tough lightweight plastic is the basis of many composite materials, such as those that make up boat hulls, tennis rackets, rocket and aircraft skins, and many other items. Each side of the area shown is 20 microns long, or roughly the size of a biological cell (2×10^{-5} meter).

Offensive Jelly It could have been devised by a sadistic medieval weaponmaker. Instead this spearheaded, multibarbed, toxin-bearing microweapon called a nematocyst is part of a box jellyfish's (*Carybdia alata*) arsenal. The nematocyst, with a trailing thread that carries poison, is shot like an explosive harpoon propelled from capsule-like structures on the jellyfish's tentacles. Jellyfish release these toxic barbs by the thousands when the creatures sense threats or that potential food is nearby. A good salvo propelled from the animal's tentacles can paralyze and even kill any hapless creature that gets too close. The nematocyst structure shown is about 20 microns in length (2×10^{-5} meter).

Bioarchitecture As far as fish go, a zebrafish isn't particularly spectacular looking. Like anything biological, however, beauty can be found down in its very cells and molecules. Here, an electron microscopist has scanned the skin of a zebrafish, revealing some amazing microanatomy. The most prominent feature is the filamentous pillar comprised of multiple fiber bundles, each emanating from a single mechanosensory hair cell. These motion-sensitive structures trigger neural impulses that keep the fish's brain informed about water currents and its own motions as it swims. The fiber bundle is 20 microns high (2×10^{-5} meter).

Allergenic Duo This close-up view of ragweed pollen reveals the beauty of the renowned allergen. Like many types of pollen, ragweed pollen boasts a breathtaking geometricity and symmetry. This particular brand of pollen is the number-one cause of hay fever symptoms in the United States. A single ragweed plant can yield over one billion pollen grains each season. The U.S. Department of Agriculture has gathered data indicating that ragweed that grows under conditions with higher carbon dioxide levels produces more pollen, so it is possible that one consequence of the higher atmospheric carbon dioxide levels associated with global warming could be more allergy misery. These two grains have been imaged with a scanning electron micrograph and then colorized with a computer. The diameter of the grains is about 20 microns (2×10^{-5} meter).

Cellular Illuminati With molecular labels that attach to specific cellular structures and then glow particular colors when excited under a fluorescence microscope, this endothelial cell from the lining of a cow's pulmonary artery (the one that shunts blood from the heart to the lungs) becomes a stunning work of light. The faint blue pattern marks the location of tubulin protein and constitutes a tiny internal skeleton. It was localized with the help of antibodies that bond specifically to tubulin molecules and that were modified with a molecule that glows blue under a fluorescence microscope. F-actin protein molecules, which are involved in cell motions such as those that enable blood vessels to dilate and contract, were stained with a red-fluorescing label. The pointillistic green staining derives from a wheat-germ-derived molecule called agglutinin to which a green-fluorescing component had been attached. Here it binds to endosomes, which are vesicles for importing proteins into cells. The endothelial cell spans roughly 20–30 microns (2–3×10^{-5} meter).

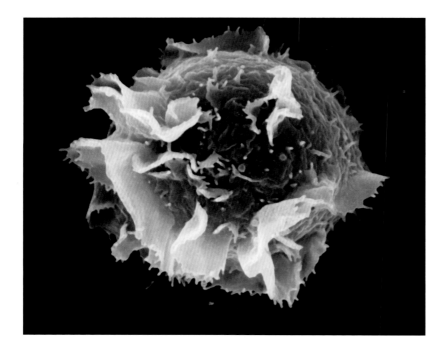

Cellular Sentinel The emergence of the immune system predates the emergence of the nervous system. In a human being, these wondrously complex systems involve a comparable number of cells. One type of immune cell, known as a mast cell, responds to the presence of threatening molecules as well as to allergens by releasing histamine, which further activates the immune system. This scanning electron micrograph reveals the beauty of the beast—this gorgeously shaped cell, with its flowing seaweed-like surface, has plenty to do with the world's totality of allergy-caused misery. The diameter of a mast cell is up to about 30 microns (3×10^{-5} meters).

Light and Mirrors In petrography, a thin section of a rock or mineral specimen is polished on both sides so that its constituent grains and internal structure can be seen under a special microscope. But it's possible to apply optical techniques to learn even more about a specimen's crystal details. One commonly used technique is to view a sample between two polarizing filters, similar to the ones used in antiglare sun glasses, oriented at 90° to each other. Polarized light, which is created after passing white light through a polarizing filter, is light whose electromagnetic waves oscillate in a specific orientation in space, such as in a specific plane. When so-called plane-polarized light passes through a second polarizer, light is completely filtered out and the scene becomes dark. But when geologist and artist Christine Skirius photographs thin crystal films, which can redirect the path of light between the polarizers, fantastic and colorful compositions show up. The details of these compositions arise from slight differences in crystal thickness and variations in the fine textures of the crystals. Skirius makes the films by melting powders of vitamins, drugs, pesticides, industrial agents, and various organic compounds on heated slides and then allowing the liquids to cool and recrystallize. She never quite knows what to expect from the process, and the results frequently are stunning. From such images, experts often can read specific details, such as the orientation of molecules within crystal domains of the film. Skirius has named the examples here (clockwise from upper left) *The Kiss, Tear Drop,* and *Tsunami*. The films are about 30 microns thick (3×10^{-5} meter).

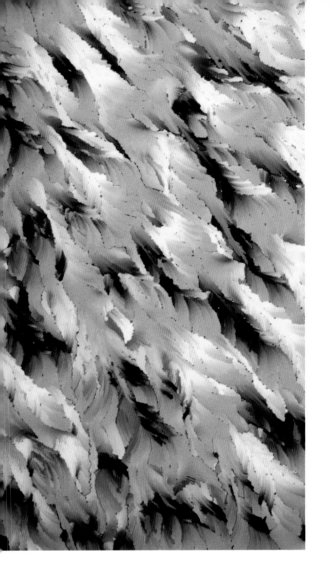

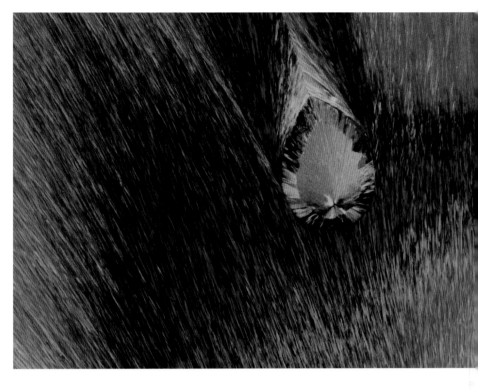

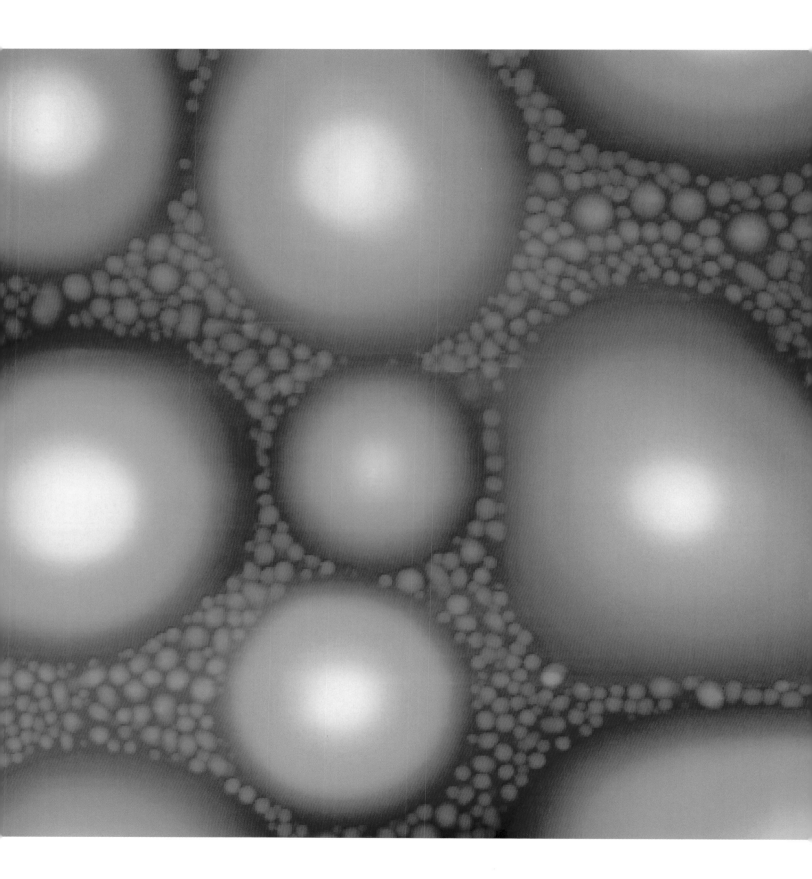

Metallic Glass Take a molten alloy and spray it into a thin sheet so that it solidifies too quickly for its constituent atoms to assemble into crystalline grains—as occurs during the normal formation of metal—and you end up with a metal whose structure is amorphous, like that of glass. Such materials can be especially strong because they lack internal boundaries where the material can come apart. They also are useful in electric applications such as transformers, which step the high voltages from power plants down to the lower voltages that power households. It turns out that metallic glasses are good at reducing and managing the heat produced in this process. Here, a tiny patch of a metallic glass alloy of aluminum, nickel, and yttrium was imaged using an atomic-force microscope. During cooling and solidifying, roughly hemispheric surface features formed in two different size classes. The smaller range from about .25 to 1.0 microns (2.5×10^{-7} meter to 10^{-6} meter), and the larger ones range from about 5 to 10 microns (5×10^{-6} meter to 10^{-5} meter). The entire image covers a square that is roughly the same size as a small biological cell, or about 30 microns on a side (3×10^{-5} meter).

(overleaf) **Neural Pollock** A human brain, when sliced thin and stained so that the otherwise-transparent cell bodies stand out, looks like an impossible cacophony of shape and connection. Yet it is in those trillions upon trillions of details that a lifetime of activity and thought is orchestrated and recorded. In this preparation, a slice of nervous tissue comprising nerve cells in gray matter was stained and then imaged with a light microscope. The cell bodies, axons, and dendrites—the fibers that interconnect the cells—combine into a rich composition every bit as arresting as a Jackson Pollock painting. The larger cell bodies span about 30 microns (3×10^{-5} meter).

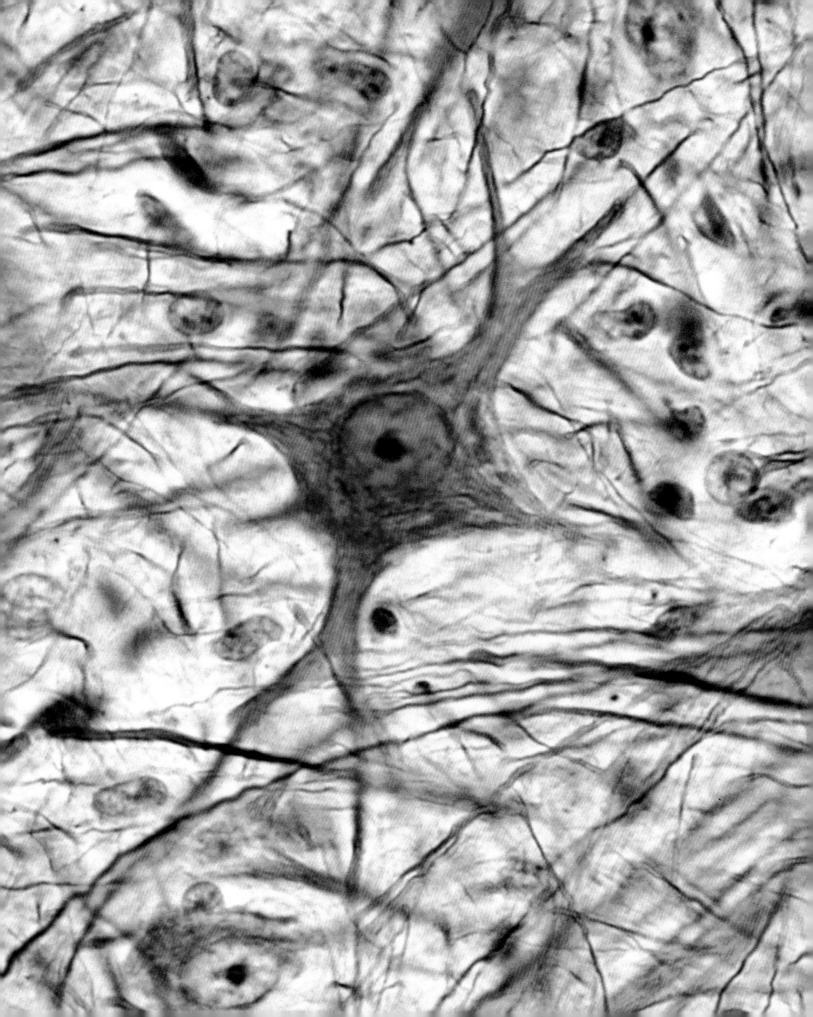

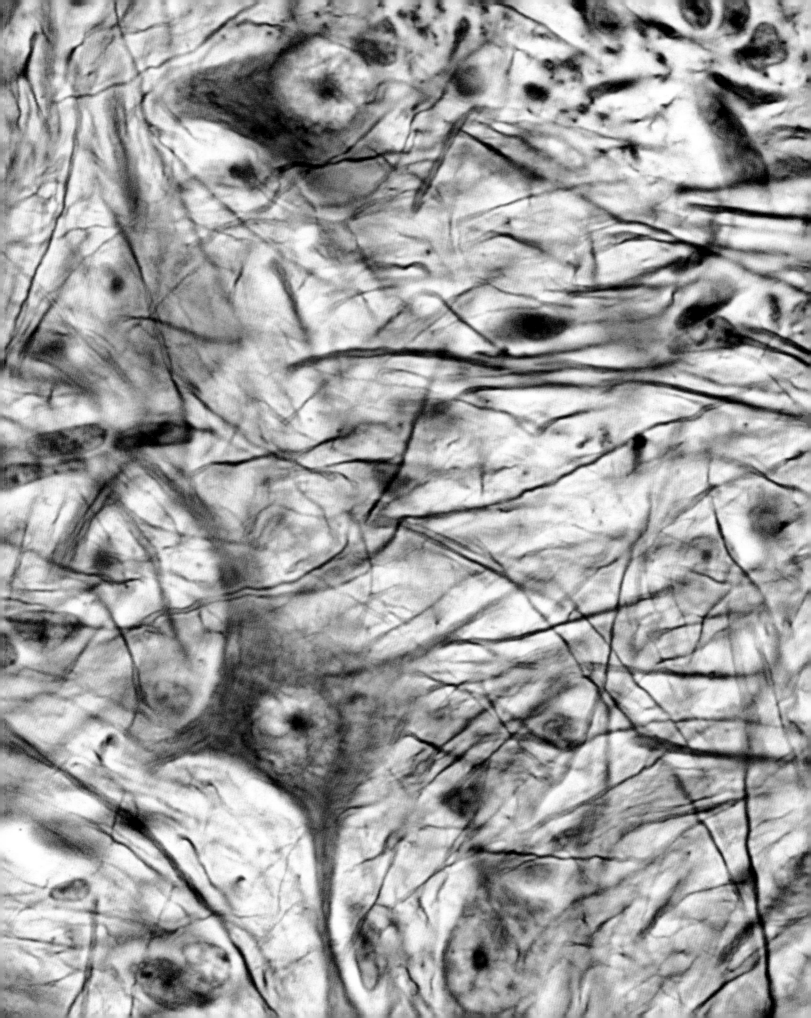

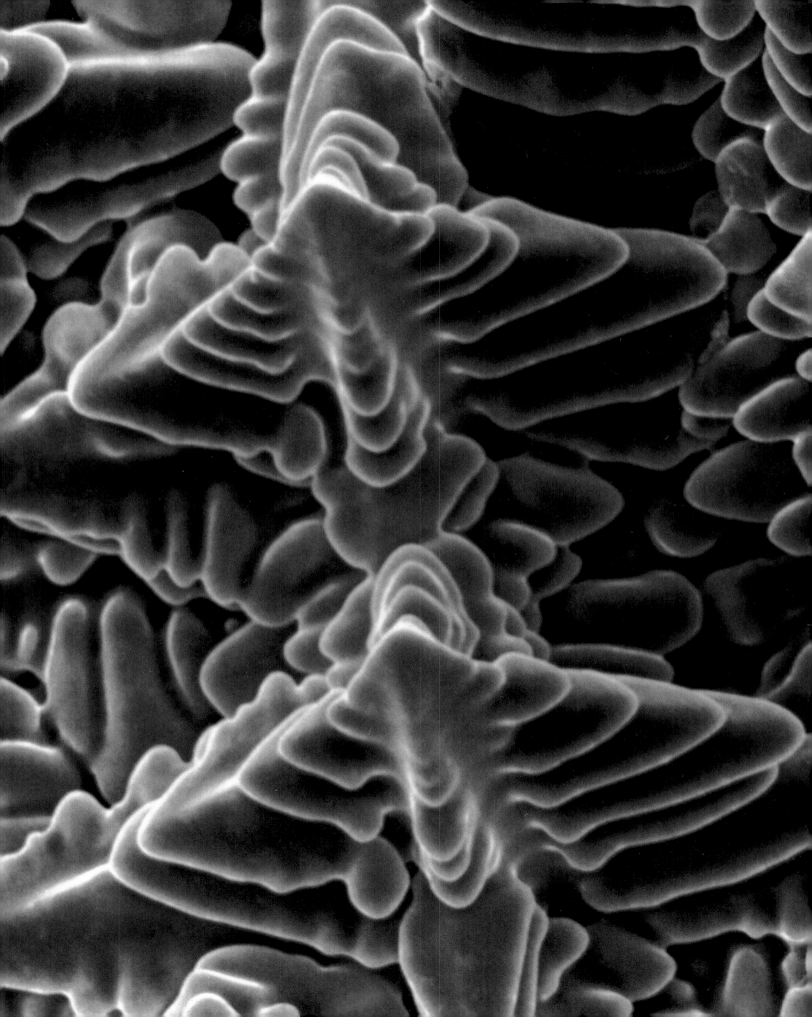

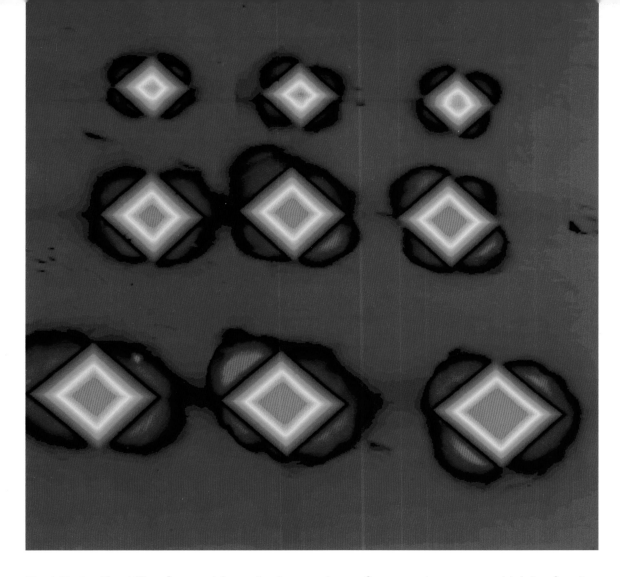

Hard Marks The ability of a material to resist damage when surfaces come into contact with it is a function of the material's hardness. Engineers often measure hardness by applying a metallic or even diamond point with precise and standardized amounts of force and then examining the resulting indentations in the material. Here, a microindenter has plunged into a thin film of gold with three different forces, three times each. The resulting indentations have been imaged with an atomic-force microscope, whose stylus-like probe could discern each indentation's various features, including the slight wellings of gold on the perimeter of the indentations, shown here in purple. Color was added after the fact. The largest indentations span about 35 microns (3.5×10^{-5} meter).

Metallic Forest It almost could be a snapshot from a helicopter hovering above a stand of snow-covered alpine trees. The tree-like structures here actually are microscopic, formed during a welding process. More specifically, a spot of a nickel-based superalloy—the kind used to make turbine blades in the engines of airliners—was heated to melting and then allowed to resolidify. As the molten metal came in contact with the cooler spots on the solidified portions of the weld, the branches of the metallic trees were formed. The result was captured with a scanning electron microscope, revealing this stand of superalloy microtrees. Each tree is about 30 microns thick toward its base (3×10^{-5} meter).

Semiconductor Pâté Among the world-changing class of semiconductors, silicon has been the superstar. But even with silicon-based microprocessors that crank at speeds of billions of calculations and more per second, electronics engineers are never satisfied—which is why they have been developing other semiconducting materials. One such material that already is in a number of high-performance electronic gadgets is the alloy silicon-germanium. Silicon-germanium transistors can switch faster than those made out of good old silicon, which translates into faster chips. The alloy usually is grown atop a silicon wafer, which has an effect roughly similar to that of stacking small grapefruits on top of big oranges. That slight misfit at the interface of the silicon substrate and the overlying silicon-germanium alloy can lead to atom-scale imperfections in the crystal known as dislocations. In one respect, they're good imperfections, because they can relieve the physical strain in the material so the material remains stable, but for use in chips, it would be best to have fewer defects. A tip of an atomic force microscope (AFM) responds to the sample's dislocations, enabling a computer to construct an image of the surface. The total area depicted could fit easily on the cut end of a human hair; each side of this AFM scan is about 40 microns long (4×10^{-5} meter).

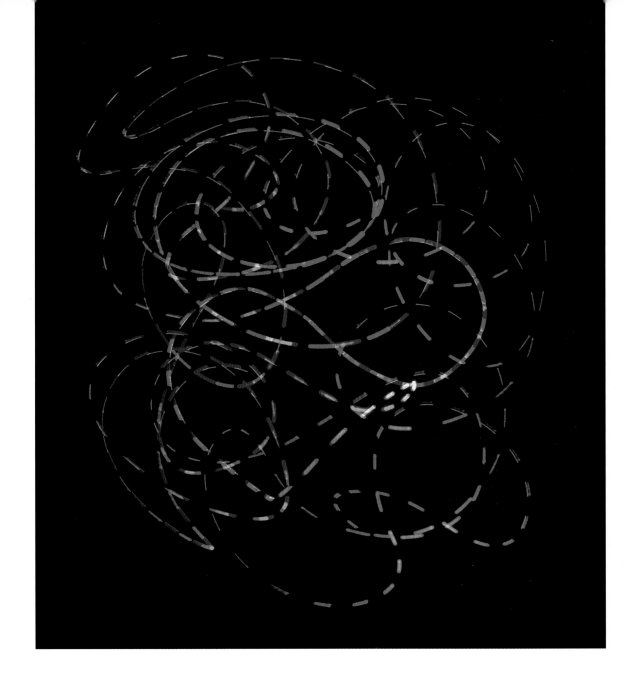

Color Talk The ongoing information revolution would have had a harder time unfolding were it not for optical fibers that confine and carry laser light the way copper wires conduct electricity. At first, optical fibers carried a single color of laser light, but researchers have learned how to simultaneously send many colors of light down the very same fiber. The technique goes by a mouthful—wavelength division multiplexing (WDM). The payoff is that WDM greatly multiplies the number of phone calls, the amount of imagery, and more generally, the volume of data, that can be transmitted down a single fiber. Rather than stringing thousands of additional miles of optical fiber to increase capacity, communications companies can upgrade their systems for WDM operation by fitting existing cables with laser and detection systems that produce and detect multiple wavelengths of light. In this image, ultrashort pulses of red, green, and blue light are caught in the act of traveling down a twisted loop of fiber. The light-carrying cores of optical fibers often are in the tens of microns range ($\sim 10^{-5}$ meter).

The Cell Begets the Cell Every second, about twenty-five million cells in an average adult divide into two daughter cells. The cell's miraculous molecular machines orchestrate the replication of DNA in the form of chromosomes, the segregation of these threadlike chromosomes to opposite sides of the dividing cells, and the pinching off of the parent cell into two new ones. In 1879 Walther Fleming observed this process and dubbed it mitosis, "the division of threads." Since then, biologists have developed ingenious ways of making this intracellular process increasingly visible. In this image, three different parts of a dividing newt cell have been specifically tagged to reveal the mechanical details of mitosis. A commercially available stain fluoresces in blue and reveals the geometry and locations of the chromosomes. A specialized antibody that binds to microtubules that are part of the spindle structure (the molecular apparatus that moves the chromosomes to opposite sides during mitosis) is adorned with fluorescent molecules that glow green under a fluorescence lamp. Another antibody—one that binds specifically to a protein called keratin in the cell's cytoskeleton (which gives the cell form and is part of its motility machinery) is adorned with a red-fluorescing tag. In this way, the actions of mitosis become visible within the larger context of the dividing cell. The length of the spindle structure, the green-fluorescing structure, is about 50 microns (5×10^{-5} meter).

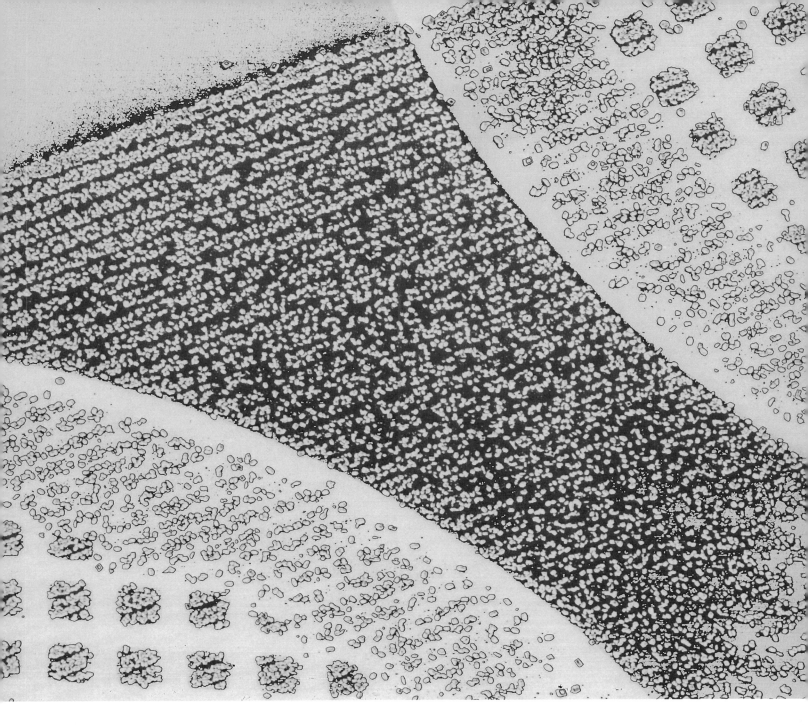

Ossified Microtechnology In a variation and extension of microlithography, the techniques used for making microchips, researchers have been finding ways to lay down microscopic patterns of many different materials. The generalized ability to work with materials at ever-smaller scales has been a theme of modern technology. In this case, scientists first deposited a single molecular layer of a substance that would later serve as a base for the growth of calcite, the stuff of chalk and seashells. But since the molecular layer was laid down using a patterned mask, the calcite also formed in the same micro pattern. Felice Frankel, artist-in-residence at the Massachusetts Institute of Technology, chose an attractive region of the much larger patterned area and then colored it with image-processing software. The square structures in the image have sides of about 50 microns (5×10^{-5} meter).

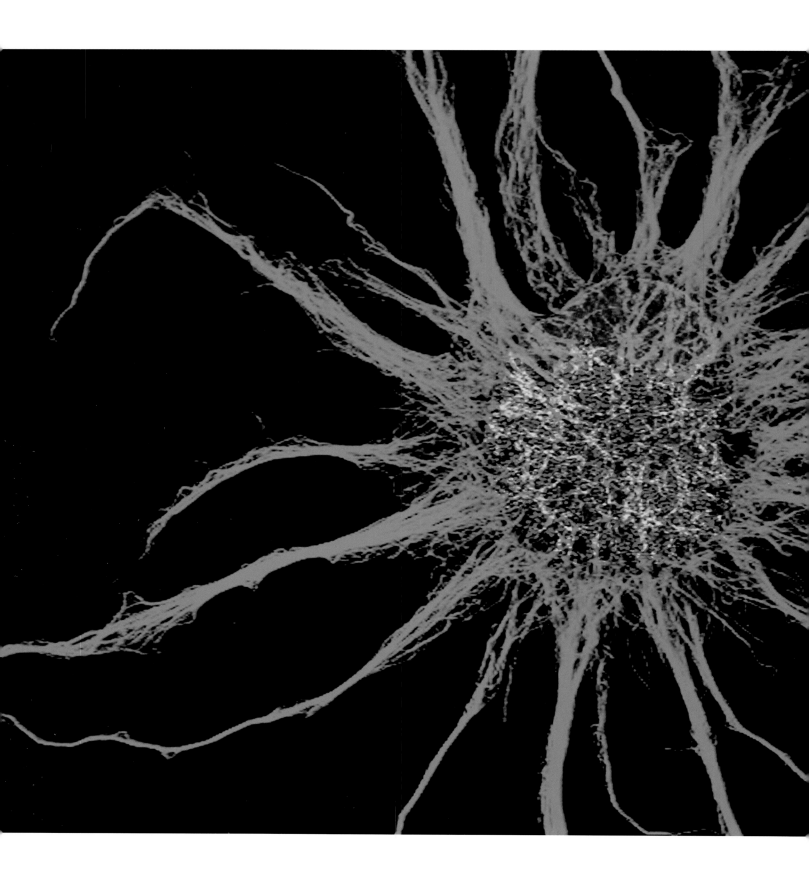

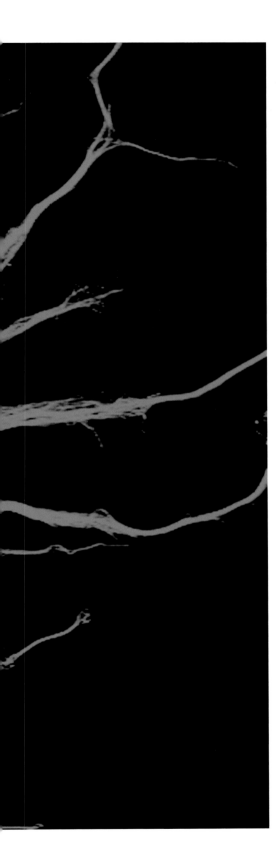

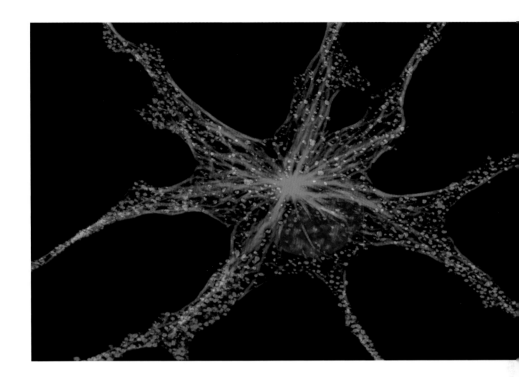

Cellular Flamboyance Once a specimen is labeled with a fluorescent dye designed to latch onto specific cellular parts, a technique known as confocal fluorescence microscopy can zero in on its very thin layers. Here, cells of a *Xenopus* frog have been prepared and imaged in this way. The cells' cytoskeletons are labeled with a green dye, and the DNA in the nuclei appears in blue. The cells' own pigment granules, which are black, look red because they are reflecting the laser light used to illuminate them. These cells are melanophores, which enable the frogs to lighten and darken their skin by aggregating the pigment granules toward the cell's center (left) or by dispersing them throughout the cell (right). The diameters of the cells' bodies are about 50 microns (5×10^{-5} meter).

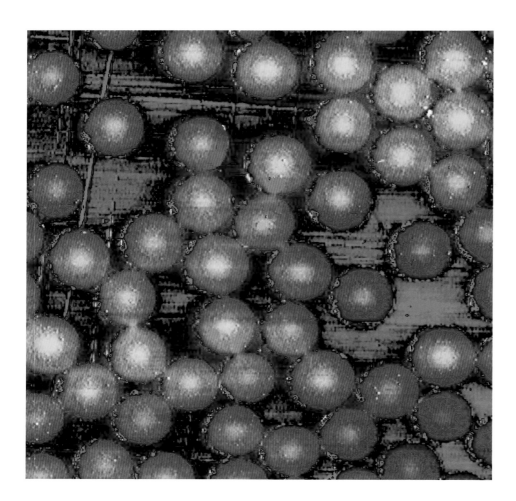

Composite Anatomy In the past few decades, investigators have been combining very different classes of substances into new material systems. That way they can combine the best properties of each substance into a new composite structure that can do more than either constituent alone. For example, adding glass fibers to plastics can lead to lightweight materials that can really take a beating. Most of the earlier uses for such composites were in high-end applications such as fighter aircraft and expensive sports equipment, but composites are now getting inexpensive enough to proliferate into many more venues throughout the constructed landscape and are taking the place of much heavier conventional materials such as steel. Shown here is an atomic-force microscope scan of a composite made of strong and tough carbon fibers (purple and orange) in a matrix of the polymer polyether ether ketone, or PEEK (green). Such images can help composite makers see, for example, where adhesion between the matrix and the reinforcing fibers might be failing. The entire scan size is about 50 microns on a side, or about the area of the cross-section of a fine hair (5×10^{-5} meter).

Weld Art The more materials scientist Lynn Boatner tried to figure out what had caused a pattern of concentric rings in a piece of melted and then resolidified stainless steel, the more mysterious it became. None of the foremost experts in welding science he queried could explain it. "That was like throwing gasoline on fire," he recalls. "We had to figure this out." Eventually, with the help of a former graduate student, Boatner did. The sample in question was part of Boatner's fundamental research into the behavior of melted metal, which is important for understanding and improving welds, among other uses. To produce a little "melt pool" in this sample, the scientists used a gas-tungsten torch. This projects a plasma in which atoms are so hot that many of their electrons wander away from their associated nuclei, leading to a hot soup of positively and negatively charged particles. This plasma exerts a pressure on the surface of the melt pool. Turn the torch off, and that pressure suddenly disappears. It's akin to suddenly removing the palm of your hand from a bucket of water on whose surface it was resting. In this case, the liquid stainless steel responded with an oscillation that froze as the metal solidified. The rings, which are close to the thickness of a human hair, are made particularly visible using a technique called interference contrast microscopy, in which two beams of light impinge on the sample. Different surface features are better or worse at reflecting specific colors that emerge from the interference interactions between the two beams. In this case, the peaks or valleys of the rings, as well as the steel's crystal structure, work together to create the tree-ring-like pattern. Each band is about 50 microns wide (5×10^{-5} meter).

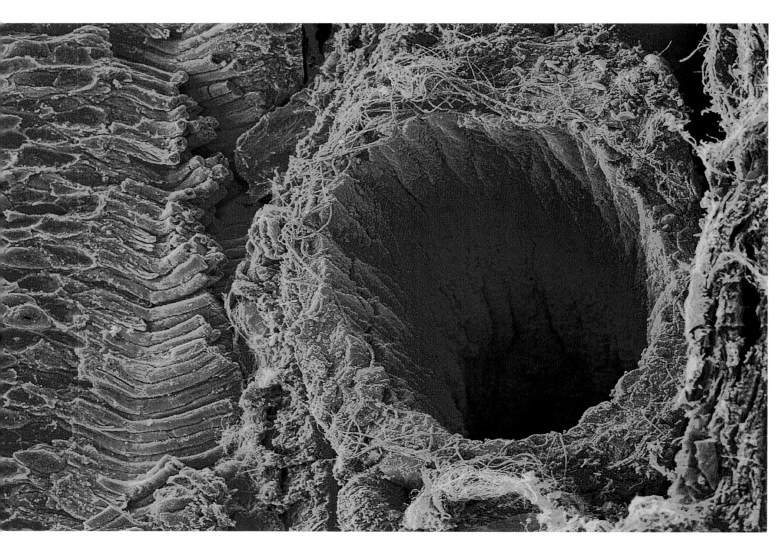

Bloody Light This formidable biological landscape was once the locus of vision for a guinea pig. At the left is a mat of rod cells, the retinal cells that capture photons and then convert that light energy into nervous impulses that the brain integrates and processes into visual information. The seemingly bottomless pit is but one microscopic vessel of a network of blood vessels that infiltrate the eye's choroid layer, which supports and nourishes the retina. The colors of this scanning electron micrograph were added later and do not reflect the actual colors of the tissue. The blood vessel spans about 50 microns (5×10^{-5} meter).

Thousand Islands Making microelectronic chips has a lot to do with taking a photograph of a pattern, usually one that has to do with circuitry, and then repeatedly transferring a miniaturized image of that pattern onto a series of silicon wafers. It's just as easy to transfer tiny images of other things instead, including, in this case, grains of sand. A miniaturized pattern of sand grains was projected onto a polymeric coating atop a wafer. In those areas of the coating that were illuminated with ultraviolet light, the polymer was strengthened, leaving behind weak areas of polymer that were then washed away to expose the underlying silicon. Then the preparation was subjected to a plasma—a hot mix of electrons and positively charged ions—that etched into the silicon wherever it was exposed. The result is a beautiful microscopic relief map that retains the outlines of the grains of sand, though not their topography. Interestingly, sand is made of silica, which is the raw starting material from which silicon wafers are fabricated. The height of the mesa structures is 50 microns, a fine hair's width (5×10^{-5} meter).

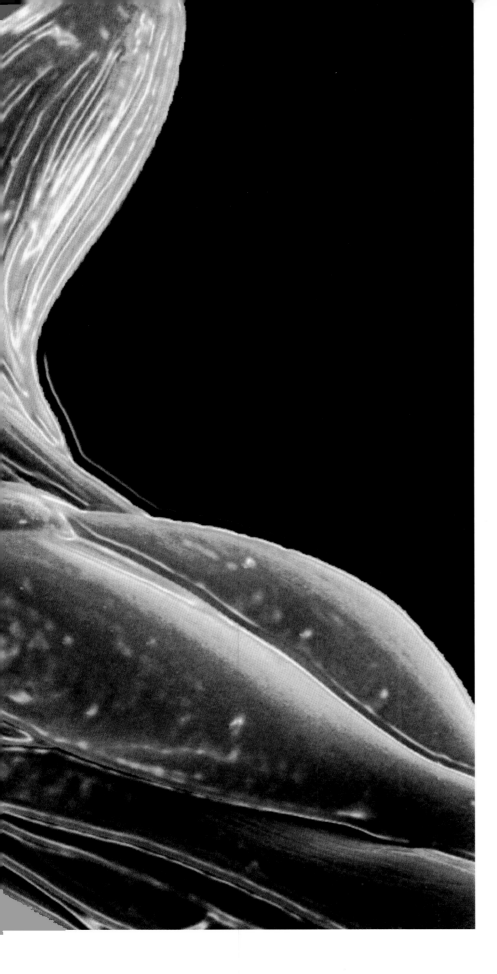

Voluptuous Seed Plants are fascinating at many spatial scales. In this case, Frieda Christie of the Royal Botanic Garden in Edinburgh used a scanning electron microscope to examine the microstructure of a seed from *Aeschynanthus tricolor*, a member of the African violet family. This specimen was collected in Borneo and cultivated in Edinburgh. There are about 160 species of *Aeschynanthus*, and the structure of their seeds, including their lumpy protuberances, or papillae, is important in their classification. When these structures are imaged with a microscope and then gently colored and processed, their beautiful forms become all the more arresting. The individual papillae range in diameter from under 50 to more than 100 microns (5×10^{-5} meter to 10^{-4} meter).

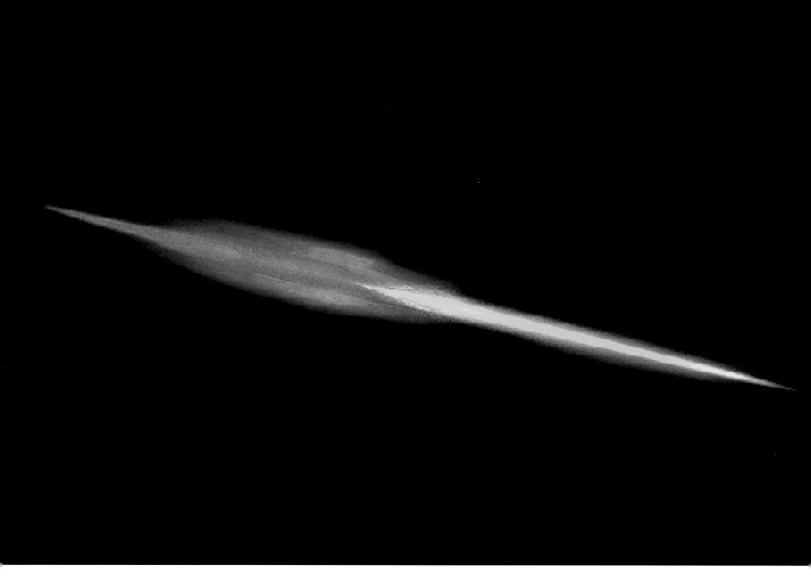

Six-Quadrillionths of a Second A lot of things in this world happen so fast that it's impossible to see them directly. This image shows what happens to a laser pulse that lasted a mere six-millionths of a billionth of a second, or six-quadrillionths of a second (6×10^{-15} second). A quadrillionth of a second is known as a femtosecond; a femtosecond is to a second what a second is to nearly 32 million years. Femtosecond pulses of light are valuable because they match or exceed the speed of such phenomena as transistors switching on microchips and chemicals undergoing reactions. With them, scientists can attain what amounts to snapshots, even movies, of molecules reacting with one another. In this image, a mix of wavelengths, or colors, in a six-femtosecond pulse of light has spread out, rendering visible the pulse's spectrum. The shorter-wavelength green light is on the leading edge while the longer-wavelength light trails behind, creating what appears to be a luminous dart. Much of modern communication is carried via countless short light pulses traveling inside optical fibers whose cores have diameters in the tens of microns range ($\sim 10^{-5}$ meter).

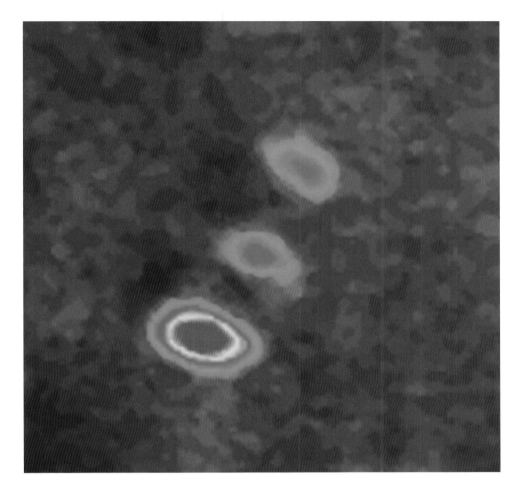

Invented Matter Since 1995, physicists have been making an exotic state of matter first predicted as a theoretical possibility in the 1920s by Satyendranath Bose and Albert Einstein. These Bose-Einstein Condensates (BECs) often consist of thousands or millions of atoms in exactly the same quantum state; the entire collective of atoms becomes a singular quantum entity that some call a superatom. Although BECs remain largely the playthings of basic science, some investigators have been looking into them as the basis for new kinds of lasers (ones that emit beams of atoms instead of light) and optical switches for communications or computing devices. This image comes from an experiment that scientists performed after using lasers to corral and cool a tiny cloud of about 35,000 rubidium atoms to a temperature barely above absolute zero. Under these conditions, the atoms underwent the expected quantum transformation into a BEC. As the BEC began expanding, the scientists imaged light scattering from it, which enabled them to measure a quantum state known as spin. They had expected that each atom in the BEC would have one of three possible values of spin. The three-lobed image confirms that prediction and indicates that there are about three times more atoms in the BEC in one state than in the other two. (The different spin states were discernible because the atoms in each state interact differently with a magnetic field and these differences can be picked up with a detector.) Understanding and controlling quantum features like spin states of atoms could lead to such devices as quantum computers that can, for example, instantly find patterns in the most massive repositories of data and information. The largest lobe of the image spans about 60 microns (6×10^{-5} meter).

Still Life with Microskeletons The sea is filled with a menagerie of microscopic creatures, many of which encase themselves in protective mineral coatings. The intricate shapes of these coatings, which often are composed of silica—the stuff of glass—have been tantalizing to scientists and to ceramic engineers who know more than anyone how difficult it is to imbue brittle ceramic with structures of such finesse and at such small scales. Shown here are three specimens of radiolarians, a type of plankton that lives in vast numbers in the oceans. These have been falsely colored to produce an enigmatic and gorgeous still life. Radiolarians have various sizes and can range in size from 50 to 300 microns (5×10^{-5} meter to 3×10^{-4} meter).

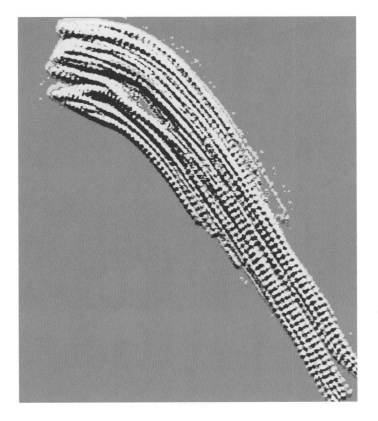

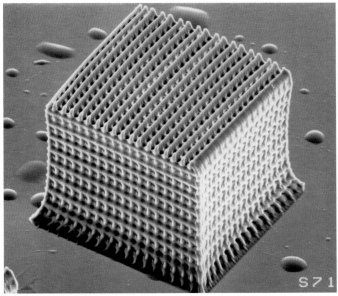

Muscle Molecules Heart muscle is astounding. Second after second, day after day, and, in our case, decade after decade, it contracts and relaxes in a steady rhythm that keeps blood flowing. Heart muscle is made of cells known as myocytes in which molecular engines made largely out of the proteins actin and myosin orchestrate the cells' contractions while structures made of cytoskeletal proteins maintain the cells' shape. In this image, biologists used enzymes to digest a rat myocyte in such a way that the cell's actin framework—its contractile engine—became prominent. Then they used a deconvolution microscope that is designed to focus in at specific depths of an already tiny object. With this tool, the scientists were able to take forty pictures through the cell at intervals of .25 microns. With the help of software, the researchers then recombined the individual images and colorized them to attractively depict the heart cell's entire actin framework. The myocyte is about 100 microns long (10^{-4} meter).

Microconstruction Using molecules that connect when simultaneously exposed to a pair of laser beams, chemists found that they could create supremely intricate structures with applications ranging from three-dimensional data storage to microscopic optical and mechanical devices. The process is a three-dimensional correlate of the fabrication methods used to make two-dimensional microcircuitry. The stack-of-logs type of construction shown in this scanning electron micrograph— which spans about the width of a thick human hair—has a combination of solid areas and voids that can control which wavelengths of radiation can enter it. Known as photonic crystals, these structures have potential uses ranging from filtering out radiation from cell phones to steering light around tiny corners in optical communications systems. Each side of this cubic structure is about 100 microns long at the base (10^{-4} meter).

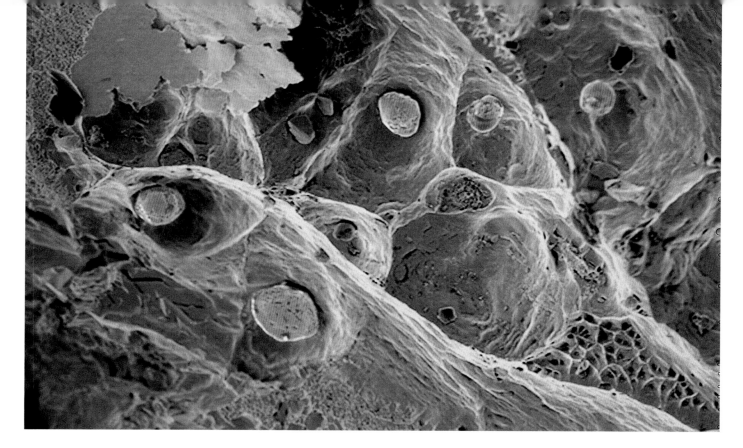

Steely Anatomy Even a few centuries ago, metalworkers would sometimes examine the fractured surfaces of metal that had failed or been torn apart deliberately by their more scientifically inclined contemporaries. Their goal was to look inside the metal, to relate its internal structure to the ingredients that went into it, to learn how those ingredients were processed into metal products, and to investigate how the metal performed. These days, microscopes are part of such fracture analysis. Researchers at the University of Oulu in Finland used a field-emission scanning electron microscope—which uses electric fields to pull a sample's electrons into a detector to supply the raw signal from which an image is constructed—to visualize the wounded metallic flesh of fractured steel. The different textures in this image are revealing. The round pink structures are inclusions—mineral or metallic junk that often weakens a piece of metal. The green flake is a bit of rust, known technically as iron oxide. The various compartments visible in the image correspond to the individual crystal grains and other microstructures that comprise the steel, which is an alloy of mostly iron and carbon, along with some smaller amounts of other elements. The colors were added to the image to highlight the different features. The area shown in the image spans about 100 microns of the sample (10^{-4} meter).

Liquid Psychedelics Falling droplets of water will thoroughly meld into an underlying cup of water, but those same droplets do utterly different things when placed within the context of a so-called nematic liquid crystal whose molecular constituents line up in the same direction, kind of like a box of straws. This type of liquid crystal can model the behavior of many materials, including biological tissues, and have many uses ranging from the development of better salad dressings to more effective blending procedures for polymer manufacturers. This light microscope image relies on polarizing filters that allow only those light waves that oscillate in specific spatial directions to get through to reveal some strange behavior in water droplets that have insinuated themselves within the liquid crystals. The droplets (the dark line of dots) arranged into a linear chain in the largest liquid crystal structure (pink) as well as in some of the other liquid crystal domains. This image was made by using a single drop of liquid crystal as a tiling unit to form the background and then superimposing a larger image of a more complicated drop on top. This big, four-armed drop spans about .1 millimeter (10^{-4} meter).

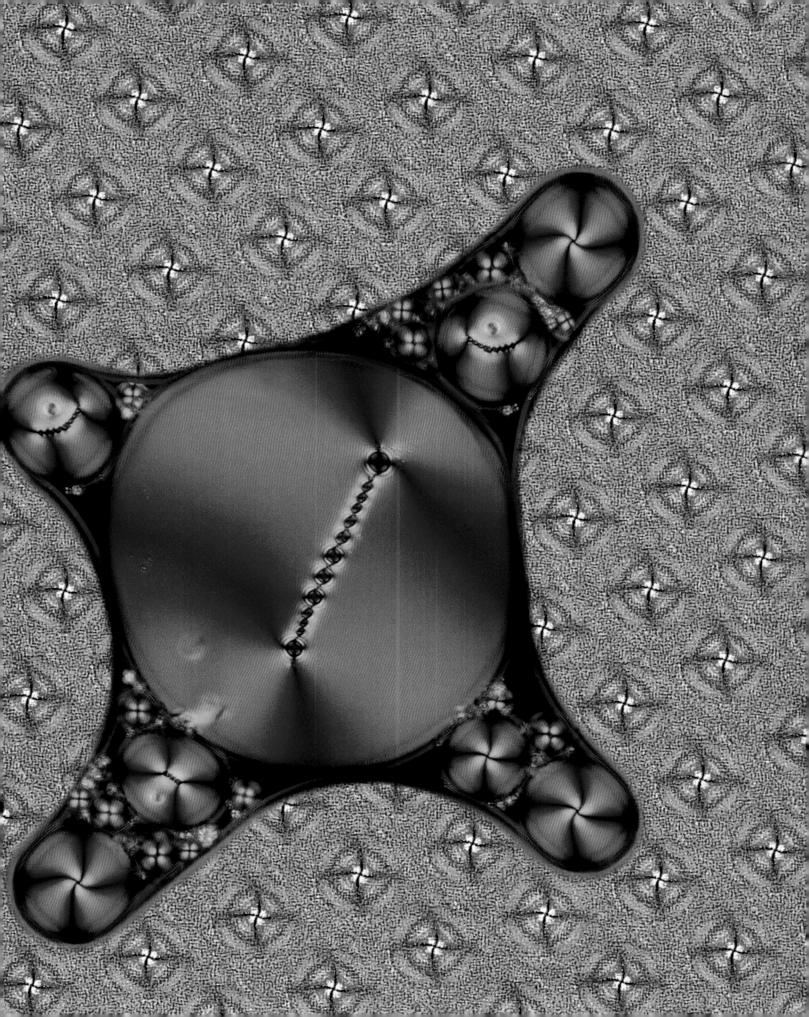

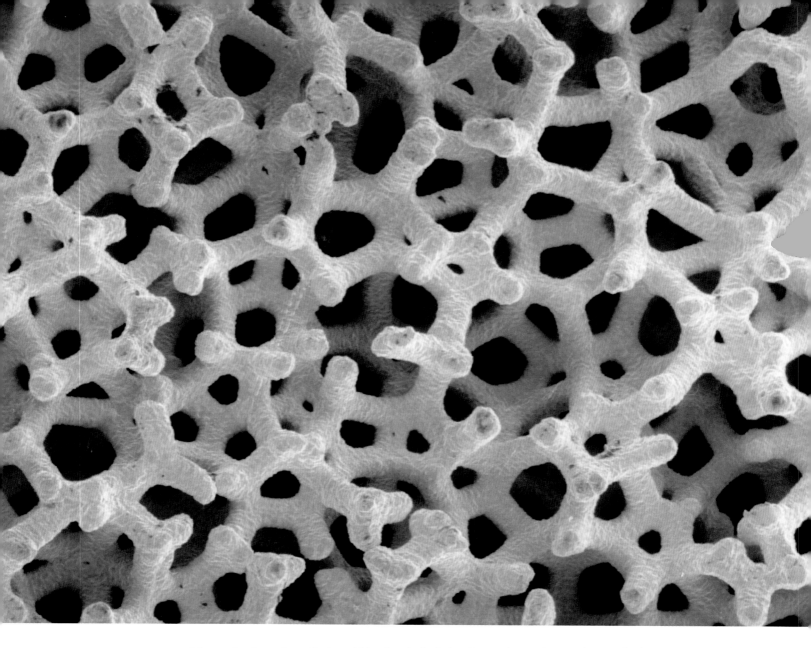

Tissue Engineering Flesh and blood and all of the tissues that make up a human body are examples of natural tissue engineering. In the medical profession's never-ending quest to reverse disease and injury and stave off death, an entirely new field of tissue engineering has emerged. Its goal is to find ways of convincing either naturally occurring cells, or ones that are modified genetically in some way, to grow into, for example, replacement bone, heart valve tissue, or skin. After all, the demand for transplant tissue always outstrips the supply. This electron microscope image shows the foam-like microstructure of what tissue engineers at the young biotech firm Cytomatrix say mimics the three-dimensional architecture of many organs and therefore provides an environment conducive to the growth of cells. The structure, a carbon framework coated with certain metals, is inert and biocompatible. By seeding it with, say, immature blood cells (including hematopoietic stem cells), the company is aiming to develop an "artificial thymus" that scientists might one day use for growing certain immune system cells in the laboratory. Such cells could then be important in cellular immunotherapies, treatments in which immune cells are injected into patients with the hope that the cells will aid, boost, or restore the patient's own immune system. The same matrix and seed cells might also be used for literally growing blood. The diameter of the framework's members in this image is about 100 microns, or just a bit thicker than an average human hair (10^{-4} meter).

Science of Stick Adhesive tape was not available until 1930, when 3M began sending its newly invented Scotch Tape to prospective customers. Now life without tape would seem downright backward. The central challenge in making tape is to coat some base material, such as paper or clear plastic, with a substance that will adhere to other materials when pressed onto them. In this microscope image of a piece of clear tape pulled away from a surface, the adhesive formed into a series of ridges as it was forced to balance its tendency to stick to a surface while being physically pulled away from it. By examining these pull-away structures, researchers can better determine what kinds of materials make the right kind of tape for different conditions. Adhesive tape typically is about as thick as a typical sheet of paper, or about 100–150 microns (10^{-4} to 1.5×10^{-4} meter).

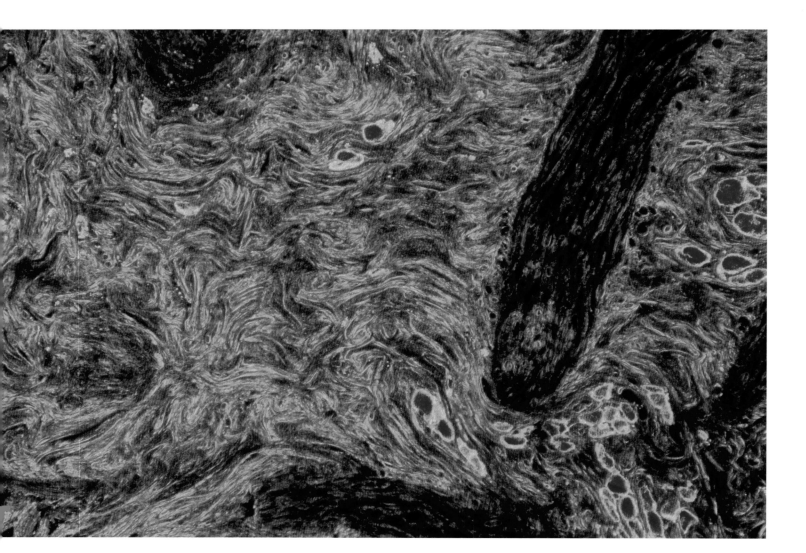

Van Gogh in Miniature The connective tissue swirls against a deep blue background, punctuated by yellow and red blotches. To photomicrographer James E. Hayden, the low-magnification microscope image evokes a comparison to Vincent van Gogh's *Starry Night*. The image is made using a section of dog skin being infiltrated by cancerous melanoma cells, which stand out here in yellow and red. The dark purple shapes correspond to strands of muscle that literally raise hairs. Under so-called Rheinberg illumination, in which white light passes through a colored filter, different aspects of biological specimens can be made more conspicuous. In this case, a central blue filter was surrounded by another filter, a yellow one. That combination yielded this accidental Van Gogh. Each melanoma cell is about 35 microns long (3.5×10^{-5} meter) and the muscle strands are about 120 microns wide (1.2×10^{-4} meter).

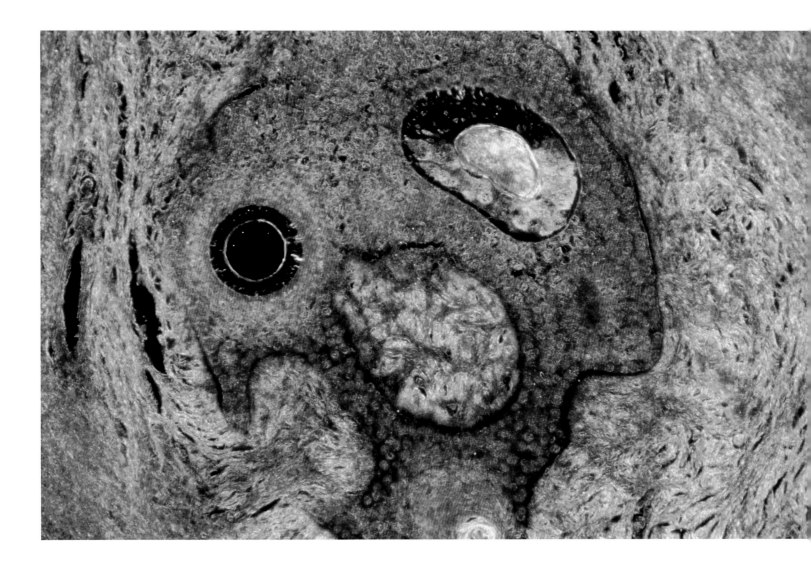

Picasso Mole When photomicrographer James E. Hayden first got hold of some histological sections of a dog's skin mole from the veterinary medicine department at the University of Pennsylvania in the early 1990s, he was instantly intrigued by the composition of forms, including, in one case, what appeared to him to be a bubble rising through the structure. Then he viewed this particular section in darkfield illumination, which brought out different features, and he saw faces everywhere. One of them has the nose pointing to the right, the top right bubble as an eye, and the middle bubble as a cheek. Another viewer reported seeing a different face—with a pointy nose on the left and black round eye. Another face is visible head on and has a macabre, almost Munchian feel to it. "I realized that I had the microscopical equivalent of a cubist painting," recalls Hayden. "Art and science never seemed closer." The black eye has a diameter of about 115 microns (1.15×10^{-4} meter, making the entire face about 700 microns across (7×10^{-4} meter).

Purkinje Tree Researchers can use a confocal microscope to take a series of crystal-clear snapshots of a minuscule sample at many different depths, and a computer can reconstruct the slices into a representation of, in this case, a Purkinje cell from the brain's cerebellum. Such a representation is fuller than is possible with a regular microscope that can view only one depth at a time. The extensive branching of Purkinje cells—whose primary roles are to orchestrate movement and maintain balance—allows them to interconnect with many others to form a fabulously rich nexus of neuronal connections. Such networks are well suited for integrating sensory information coming in from many locations of the body as well as for orchestrating the movement of the many muscles required for coordinated movement. Shown here is an image of a Purkinje cell from a nine-day old rat. Such cells are large; their bushy extensions can span 200 microns or more (2×10^{-4} meter).

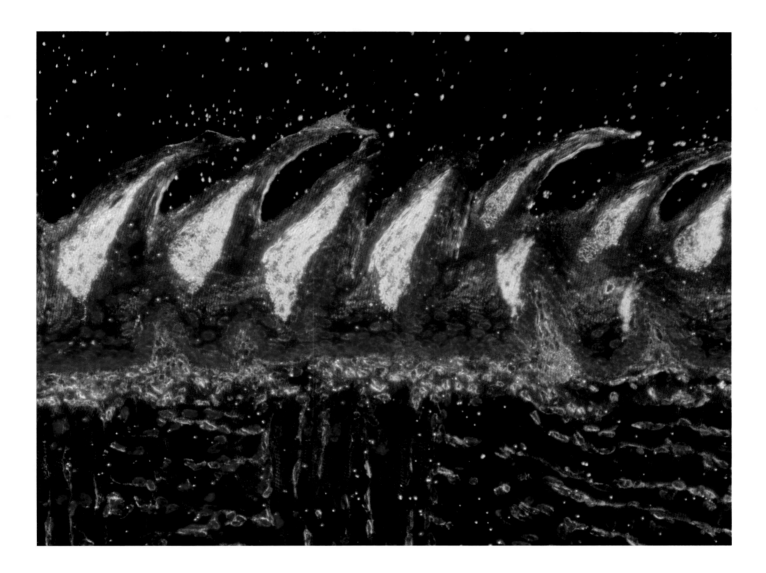

Tongue Out of Cheek People look better under certain lighting conditions than they do under others, and it's the same with microscopic preparations of biological tissues. Under the normal bright light of an optical microscope, this section of a mouse tongue looked like a montage of shapes in subtly differing shades of blue. When viewed with the indirect light of darkfield illumination, however, this specimen lashed out like a painting in a spectrum of colors. The effect was caused by light bending into different wavelengths as it passed through different tongue structures. The filiform papillae—small raspy projections that help hold food onto the tongue—are made of keratin, the same stuff as fingernails and hooves, but different areas have different densities. The yellow areas stand out because they are less densely keratinized. The ordered red and green striations of tissue at the bottom of the image are muscle fibers. The papillae are about 200 microns high (2×10^{-4} meter).

Secret Garden We most often think of plants as being anchored in the ground, seabed, or some other Earth-based floor. The species *Tillandsia usneoides*, known more commonly as Spanish moss, frequently can be seen hanging from trees and telephone cables from Virginia to Texas in the United States and as far south as Argentina. Because it lives its life anchored to other things, this class of plants goes by the name epiphyte, which combines the words for plant and "on top of." Spanish moss has no roots and absorbs water via its so-called scalehairs, which look like ribbed flower petals in this scanning electron microscope image. The microscopist Raija Peura says her choice of colors for the scalehairs was inspired by Van Gogh's *Sunflowers*. Each scalehair is about 200 microns across (2×10^{-4} meter).

Skin Teeth A shark's jagged, meat-tearing teeth are far outnumbered by the smaller denticles that cover its skin. Also known as placoid scales, denticles are bony, spiny projections with a tough, enamel-like surface. They slant toward the tail and sum into a highly engineered, hydrodynamic surface that helps direct the flow of water along the shark's torso, thereby reducing energy- and speed-sapping drag. Stroke a shark from its head toward its tail and it feels smooth. Go the other way, however, and it feels like sandpaper. The micrograph is of a thin section of a dogfish shark's placoid scales that was fluorescently illuminated in such a way that the underlying skin appeared green, and the denticles' interiors and exteriors showed as red and bluish, respectively. The image was captured not with film but digitally using a semiconductor detector, the way an electronic camera does. Each scale is about 250 microns wide (2.5×10^{-4} meter).

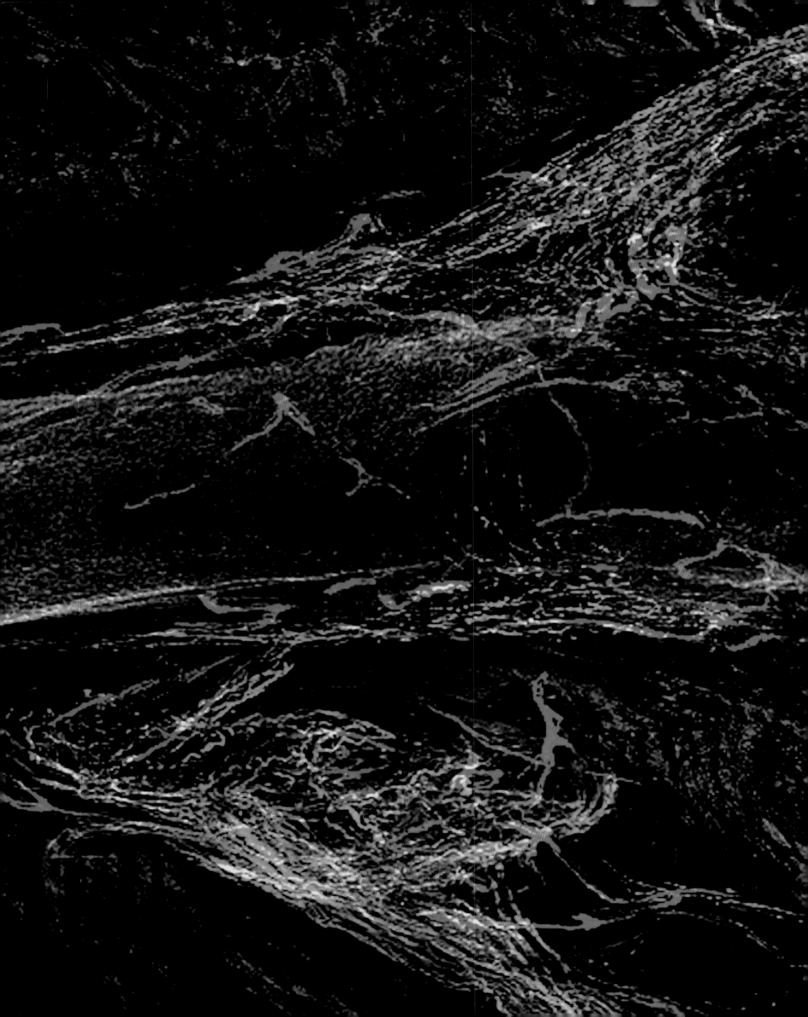

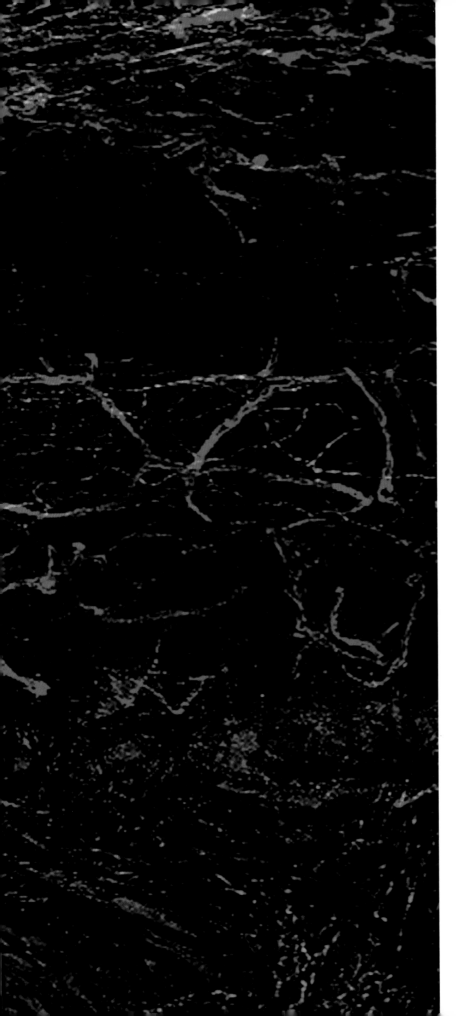

Hair Down Under At the base of each shaft of hair is a follicle, a remarkable structure from which the hair sprouts and grows. The follicle is fed by tiny blood vessels and even has nerves, which relay pain signals to more central neural centers and orchestrate subtle movements of hair and skin, the kind that raise goose bumps.

This image of a scalp hair follicle from a forty-something male is "like a mini-CAT scan," in the words of Marna Ericson, the dermatology and microscopy expert who took great pains to create it. The technique she used goes by the name epi-fluorescence laser scanning confocal microscopy. This preparation involved several different stains. One was based on an antibody that binds to nerves, which show up here as the filamentous green structures. The blood vessels, in red, were stained with a plant compound known as a lectin, to which the fluorescent compound fluorescein had been chemically attached. A third, less visible, stain was used to label a specific nervous system compound that mediates pain and inflammation processes. This image, a composite of 100 separate views, uses a process that creates images that portray depth with more fidelity than the single views of the much thinner samples that come from conventional microscopy.

The thickness of the sample was 250 microns (2.5×10^{-4} meter).

Magnetic Personality Inside some of those cylinder-shaped transformers attached to power poles is a kind of material known as metallic glass. When made with iron, boron, carbon, and a few other alloying ingredients, the metallic glass can harbor magnetism in a way that enables transformers to reduce high voltages from the power plant to the more manageable household voltages while dissipating less of the electricity as waste heat than do more conventional magnetic materials. In the image, researchers examined a piece of magnetic metallic glass with a tool known by the acronym SEMPA, which stands for scanning electron microscope with polarization analysis. In this case, polarization refers to the way in which electrons coming off the material spin. Since those spins have everything to do with the strength and direction of the material's magnetic field, this tool enabled the scientists to visualize the tiny variations in their sample's magnetism. The broad yellow, blue, and purple stripes correspond to tiny regions of the ribbon with relatively homogeneous magnetic fields. The central whirl reveals a defect in the material, a dimple that affects the properties of the magnetic field there. The zebra-striped regions represent parts of the ribbon that had melted somewhat and then recrystallized to create regions with an unwelcome finer scale of varying magnetic domains. Enough defects like these and you can end up with a transformer that hums and wastes energy. The fatter stripes in the image span about 250 microns (2.5×10^{-4} meter).

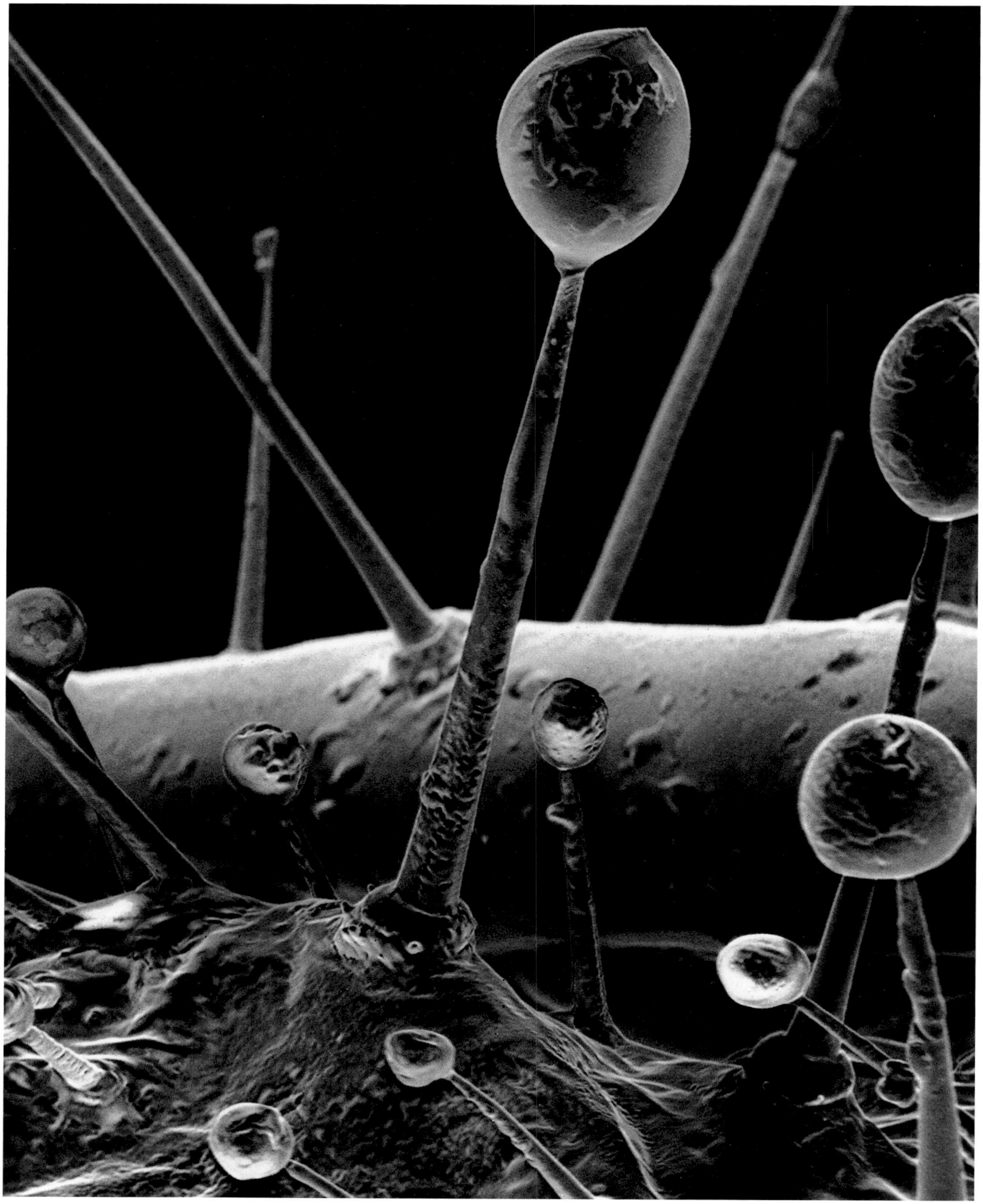

Rococo à la Mite Spanning only a few widths of a human hair when full grown, species of Cosmochthonius mites stand out for their ornate forms. They normally dwell in dry soils, but they also do fine in dry micro-environments in otherwise very wet places; this particular mite, photographed with a scanning electron microscope, was scooped out of a dry tree hole in a rainforest. Mites are important players in many ecosystems. A handful of forest humus could have populations of 100 different species of mites all going about their different and often interrelated business, according to David Walter, the mite expert at the University of Queensland, who took this portrait as well many others from the world's vast mite menagerie. Estimates of the world's total number of mite species, most not yet identified, often reach into the hundreds of thousands. This specimen, a member of the suborder Oribatida, is probably is about 400 microns long (4×10^{-4} meter).

Chemical Defense The pupa of the ladybug species *Epilachna borealis* would be far less likely to survive in the wild were it not for its ability to literally mix and match three closely related molecular blocks of its own making into a cocktail of chemicals, known generally as alkaloids. This toxic brew protects the pupa from predation. In this scanning electron microscope image, droplets of the defensive concoction glisten at the tips of the pupa's glandular hairs. Ants making contact with the fluid instantly rear back and frantically wipe the secretion off their appendages. Each hair is about as long as a folded piece of paper is thick. The adult beetles, also known as squash beetles (one of the few ladybug species that are agricultural pests), are orange with black spots. The pupa's hairs are about 275 microns long (2.75×10^{-4} meter).

The Outside of an Inside Surface Like skin, the lining of the upper vagina is always renewing itself. As the broad, flat cells at the surface slough off, new ones underneath take their places. Fine ridges mark the boundaries between individual cells while arrangements of several adjacent cells comprise the flake-like structures visible in this color-enhanced scanning electron micrograph. The four red dots mark the location of individual red blood cells resting atop the lining. The area shown spans about 400 micrometers (4×10^{-4} meter).

A Choreography of Calcium Besides its well-known role in the health of bones and teeth, calcium is a central chemical cog in muscle action. Muscle cells maintain reservoirs of positively charged calcium ions. When nervous impulses arrive at a muscle cell, the cell releases some of its stored calcium ions, which starts mechanical motion in the molecular machinery underlying muscle action. The images on this page reveal changing calcium levels at .55-second intervals within a single elongated muscle cell from the lower region of the small intestine. The calcium in particular is made visible with the help of a special dye that emits light only in the presence of calcium ions. The colors map out the concentration of calcium, with pink marking low calcium levels, and blue, yellow, green, and red marking progressively higher amounts. It was possible to image a specific portion of the cell by using a confocal microscope, which employs a laser to illuminate only a specific depth or "slice" of a cell while additional components of the instrument reject out-of-focus information from other planes of the sample. It also is possible to store images of different depths in a computer, which then can reconstruct three-dimensional microscopic portraits of samples. This muscle cell is about 500 microns long (5×10^{-4} meter).

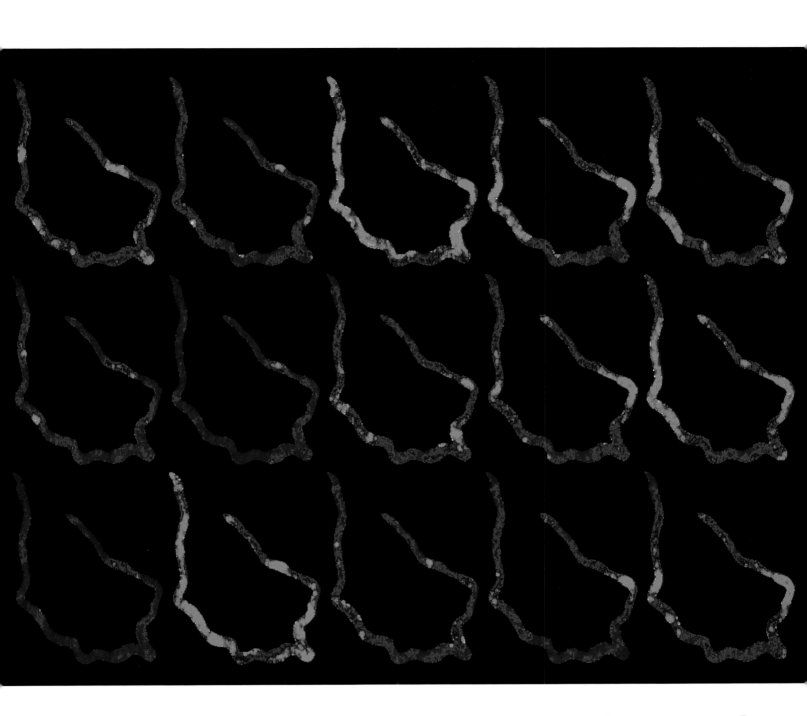

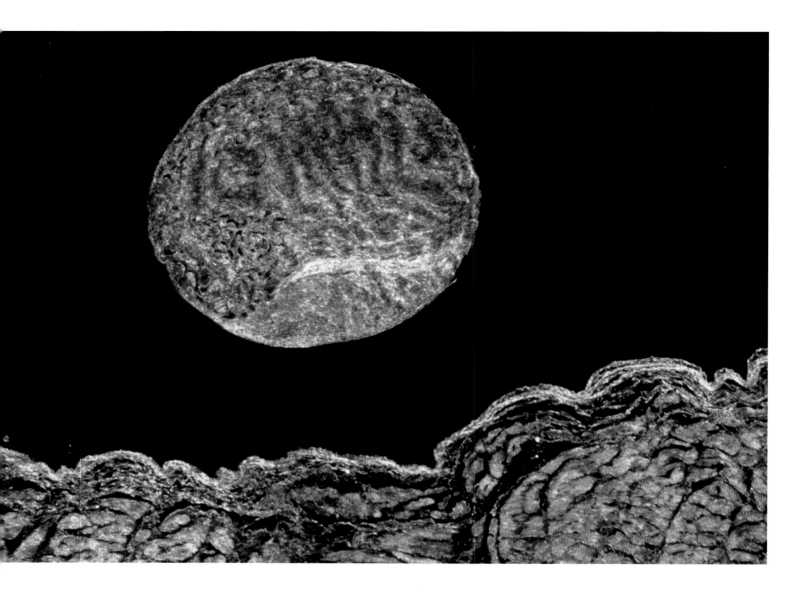

Heartrise The heart is a display of biological plumbing par excellence. An orchestration of muscles shunts blood into this vital pump, pushes it out of the pump, and controls valves to make sure there isn't much blood going in the wrong direction. Shown here in darkfield mode—in which light is blocked from passing directly through the specimen and is instead refracted, or bent, onto the specimen from the edges of a condenser lens—is a small section of a dog's heart (bottom) near where it attaches to the chordae tendineae. This strand-like structure, round in cross-section, is made of muscle (red) and connective tissue (blue) and connects a tricuspid valve's leaf-like flap to the inside wall of the heart's left ventricle, from which oxygenated blood is pumped out to the rest of the body. In this microscope preparation, the tissue is reminiscent of a strange moon rising over an equally strange planet. The photographer, James Hayden, quips that he can even see South America in the round chordae tendineae, which, to the naked eye, is about the size of the period at the end of this sentence. The structure's actual average diameter about 500 microns (5×10^{-4} meter).

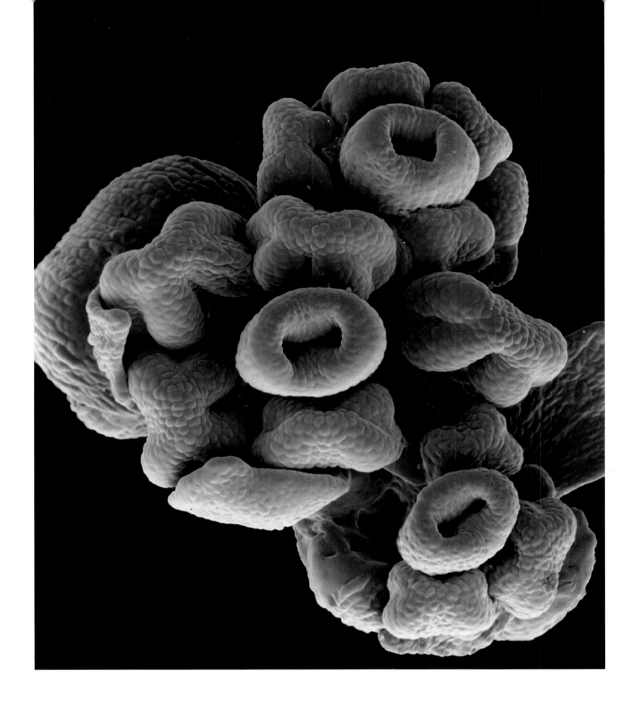

Mutant Microflowers The plant species *Arabidopsis thaliana*, a small weed in the mustard family, is to plant scientists what rats and mice are to those who study animals. It's one of the most thoroughly studied members of the plant kingdom. Like many plants, *Arabidopsis* normally does not produce flowers at the tips of its shoots, but instead continues to grow because its Terminal Flower 1 (TFL1) gene produces a protein that initiates a biochemical orchestration that results in continued growth. The scanning electron microscope image, taken by scientists at the John Innes Centre, a biotechnology research firm in Norwich, England, reveals what happens when TFL1 has mutated or become inactive: a cluster of mutant *Arabidopsis* flowers has bloomed. The individual blossoms are too tiny to be appreciated by the naked eye; the entire cluster is about the size of a poppy seed. Studies of the genetic control of plant features, such as flowering and growth, can lead to new high-tech crops and agricultural products. Fears about how genetically modified organisms might affect the environment, however, have made this area of applied genetics research one of the most hotly debated of our time. The blossom cluster spans about 500 microns (5×10^{-4} meter).

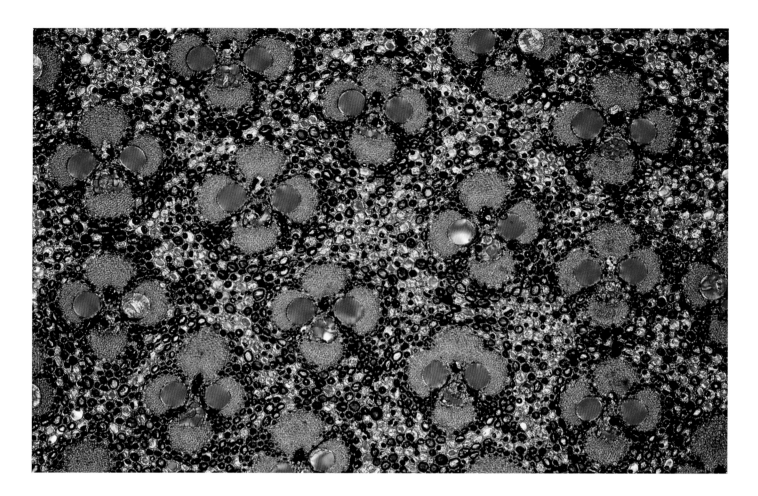

Bamboo by Accident Materials scientist Ron Sturm was using pieces of bamboo as convenient spacers between some specimens of shale, a slate-like mineral, which he was preparing for mineralogical study. One technique he used for this study involved making thin sections of the shale, which were then photographed using a polarized-light microscope, which helps reveal crystal grains and other distinguishing details of microscopic specimens. The views of the shale were interesting enough, but Sturm realized that the bamboo itself was also well worth examining under the microscope. When he made thin sections of it, he was able to reveal structures that reflect the basis of its extraordinary strength. Like reinforcing steel bars in concrete, strong and tough fibers of cellulose—arranged in quartets that look like flowers in this cross-section—course through the cellular network that makes up the bulk of the bamboo. Each four-leaf-clover structure is about 500 microns across (5×10^{-4} meter).

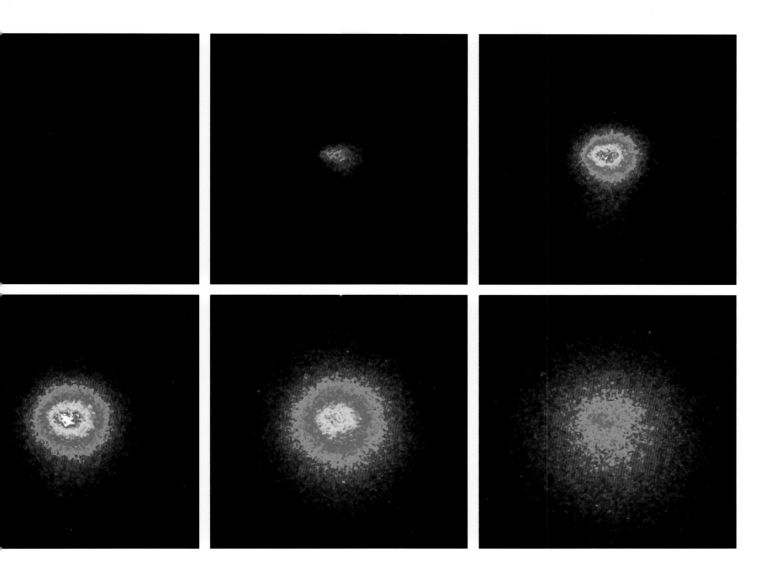

Watching Retinal Chemistry Microscopists began using dyes in the nineteenth century to highlight particular aspects of their specimens. One recent high-tech development has been to use dyes that emit light only under specific conditions or only when cells are doing certain things. In this image, a neuroscientist has used an ingenious biochemical system in which the molecules luciferin and luciferase emit light when activated by the cellular fuel-chemical ATP. (The ATP-luciferin-luciferase system evolved in fireflies and is used by these insects to generate light flashes.) This series of images reveals how the release of ATP propagates from a central point of stimulation in a retina. Red indicates the highest levels of ATP, and blue indicates the lowest. In a sense, this dye system enables scientists to watch retinal cells signal each other. The width of each image is nearly 500 microns and includes the activity of many retinal cells ($\sim5 \times 10^{-4}$ meter).

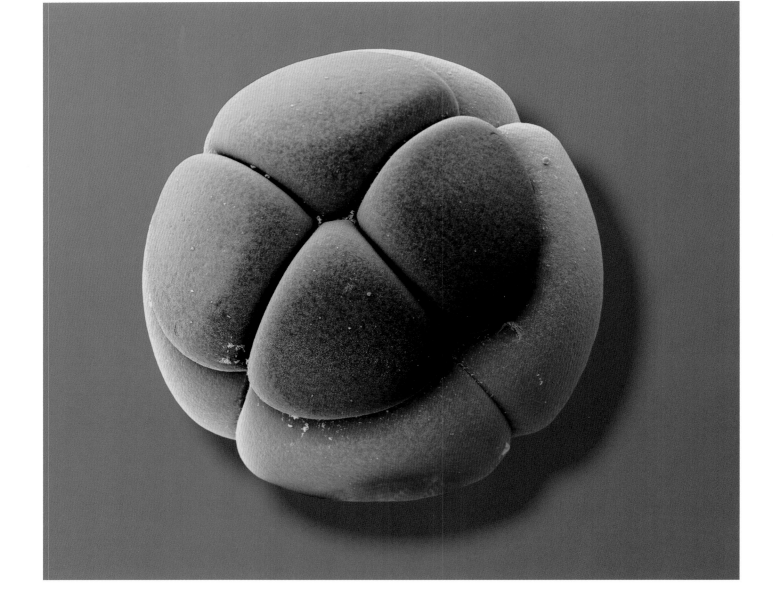

Lifeball Life begins with deceptive simplicity. A bacterial cell divides. An egg cell is fertilized by a sperm cell. But then, for multicellular creatures, that beginning unfolds into a process of varying degrees of complexity. This scanning electron microscope image shows a *Xenopus* frog embryo after the fertilized egg that it came from had undergone exactly three cell divisions on its way to tadpolehood. One cell became two, two became four, and four became eight. During development, this multiplication process goes on while natural editing processes remove some cells that were required transiently, perhaps only as scaffolding. In addition, cells differentiate into different types and different tissues as various genes turn on and off via a timing protocol whose mechanisms have yet to be revealed in detail. The embryo is 1 millimeter in diameter (10^{-3} meter).

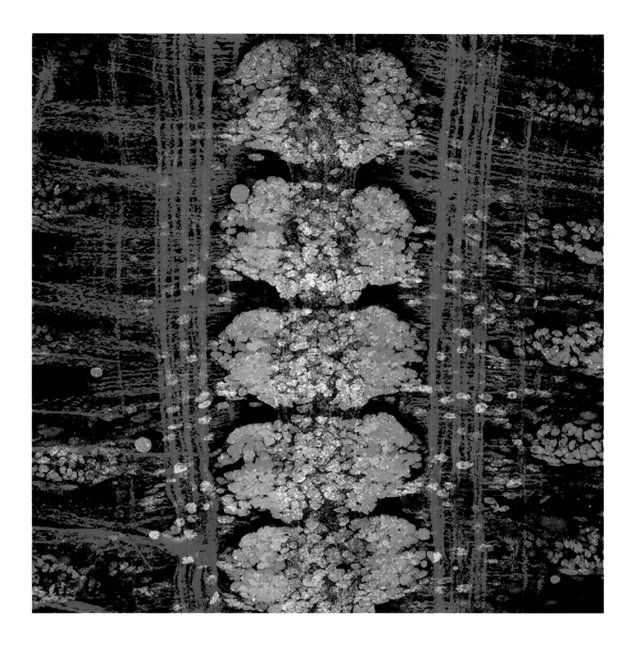

Cute Little Leech One of the great miracles of biology is that an organism's entire body—from the finest invisible cellular details to the most obvious external traits—emerges over a course of development that begins with one cell, a fertilized egg. How that cell differentiates into the many cell types that comprise a body is a subject of great fascination and a huge amount of research. Researchers have invented a molecular tracer that when injected into a specific cell of an embryo will leave glowing molecules behind in all the cells that develop from that initial cell. Here, developmental biologists injected a single so-called neuroectodermal cell of a leech embryo with a blue tracer, revealing that the cell differentiated into multicellular clusters of neural assemblages known as ganglia. Muscle fibers associated with the ganglia are visible as red, and a green dye makes cellular nuclei visible. The ganglia are in the vicinity of 1 millimeter wide (10^{-3} meter).

Laminar Lightwork When light enters a phyllo-like crystal of potassium niobium oxide like this one, it enters a multilevel framework of surfaces alternating with empty spaces. The surface of the crystal has a texture akin to crumpled paper, caused by water vapor that has insinuated itself into the tiny crevices between the crystals' layers. For light, the material becomes a complex hall of mirrors. As the light penetrates and then reflects from the many internal surfaces, it bends and mixes and interferes with itself in a thousand different ways. Add to this the reflections from the surface, and you end up with a dazzling portrait of color and bands that seem more the product of an artist with a palette than a materials scientist with a crystal and a microscope. Potassium niobium oxide falls into a category of so-called ferroelectric materials whose electrical conductivity changes in the presence of a magnetic field. That makes this class of materials useful for applications ranging from magnetically tripped switches to data-storage devices. The area depicted in the image is just over 1 millimeter wide (1.1×10^{-3} meter)

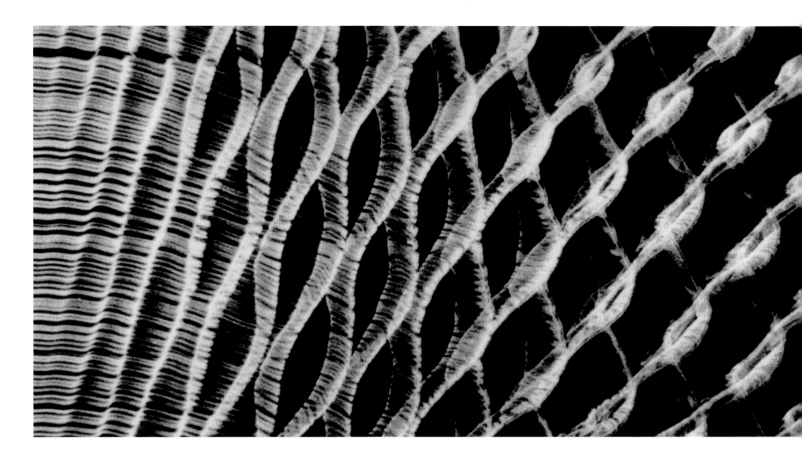

Ripples Fantastique The surface of an ocean, lake, bay, or other body of water is a cornucopia of periodic undulating activity. Little ripples play on the ups and downs of bigger ones, which themselves dance on the hills and valleys of yet-larger waves. Disturb that liquid surface in some way, say, by dropping a pebble in it or by paddling through it with a kayak, and the situation gets even more complex: Wakes from those disturbances interact with the existing pattern of ripples in ways that evolve as the wakes travel away from their source. To study such phenomena, some scientists create model systems that let them visualize how fluid flows change and evolve under different circumstances. Here, smoke generated from oil on a heated wire joins a stream of air traveling from left to right and over a cylinder (not seen in the image), generating vortices. In effect, the smoke makes the invisible flows of air visible. At first, the cylinder causes the airflow to organize into a regular pattern of vortices, which behave like waves. But downstream, these waves interact—in the central region of the image—with very small disturbances in the airstream. That interaction evolves into the wickerlike wave pattern on the right, what scientists in this field call an oblique resonance wave. Such studies can uncover nuances of fluid flow that ultimately can become applicable in venues as varied as liquid-based manufacturing and studies of pollution distribution. The upstream waves (on the left) have a spacing about 2 millimeters (2×10^{-3} meter).

Snow Safari It is not a simple matter to view snowflakes through a microscope. After all, such instruments normally operate at room temperature, and snowflakes become tiny puddles of water very quickly. Researchers have overcome this problem in various ways, including simply taking a microscope outside on snowy days. But no one has taken the art of photographing snow crystals further than William Wergin, a researcher with the U.S. Department of Agriculture, and his colleagues. First, they allow snow to fall on prechilled copper plates coated with a special cold-working adhesive ordinarily used for viewing biological tissues. That done, the plate is thrust into a vessel containing liquid nitrogen at −196° C, thereby "locking" the fragile microstructure of ice into place. Next, the samples are gently coated with a film of platinum or a gold-platinum combination. This coating makes the snowflakes far more visible through a special electron microscope whose viewing chamber can be maintained at about −185° C. Over the years, Wergin and his coworkers have amassed a stunning gallery of snowflake images. What strikes Wergin as much as anything is that the snowflakes he observes rarely display the perfect symmetry that we grow up expecting of them. But it is in the subtle asymmetries that much of the flakes' beauty resides. The diameters of the snowflakes in this ensemble, shown at various degrees of magnification, are in the 1–2 millimeter range (1–2 x 10^{-3} meter).

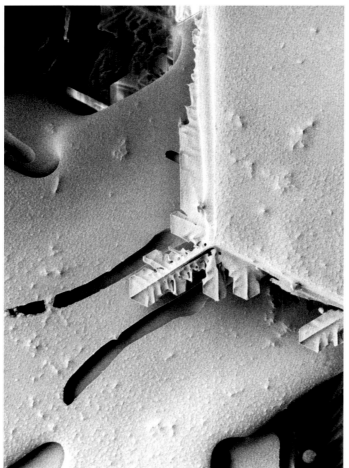

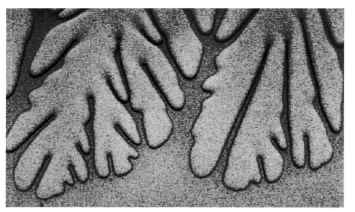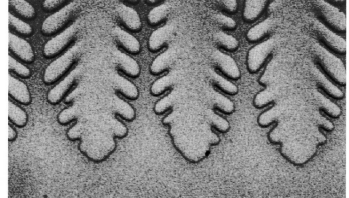

Impersonating Metal Finding ways to control the solidification of molten metal alloys into solids is a pathway to making metal that is stronger, more resistant to failure, and otherwise better than existing metals. Some researchers use a polymer—in this case poly(ethylene oxide)—dissolved in an organic solvent (succinonitrile) as an experimentally convenient stand-in for solidifying molten metal. Here, they placed a thin layer of such a polymer solution between two glass slides. By making one side of the slide warm enough to keep the material in liquid form and the other side cooler, the scientists created conditions for what is called directional solidification, an especially ordered type of crystal growth that is important, for example, in the making of the strong and tough superalloy vanes and blades used in jet engines. The different branching forms in the two images—one of them pinecone-like (right) and one of them less regular (left)—reflect the different microscopic orientations of the solidifying molecules. Just as it is possible to place a cereal box down on a table on either its bottom or on one of its longer sides, so is it possible for a crystal to lay down on a surface with a specific orientation. In the figure on the left, the crystal grows with an orientation that creates structures reminiscent of seaweed. In the figure on the right, another orientation has led to a more regular pattern of structures known as dendrites. By using a special contrast technique for imaging, known as Hoffman modulation, the researchers were able to render visible the solidification structures in what otherwise would have appeared as a transparent, almost structureless sample. The width of the images is about 1.5 millimeters (1.5×10^{-3} m).

Dancing with a Minuscule Animal The Hydra of mythology is a large, many-headed monster, but a microscale counterpart in the Cnidaria phylum shows a stunning grace in this colorized electron micrograph image. This little creature, about the size of a sesame seed, lives in fresh water. It is shown here perched atop an aquatic plant against a background falsely colored blue to reflect the organism's watery domicile. The Cnidaria phylum encompasses about nine thousand species, including those of corals, jellyfishes, and anemones. This hydroid Cnidarian is about 2 millimeters long (2×10^{-3} meter).

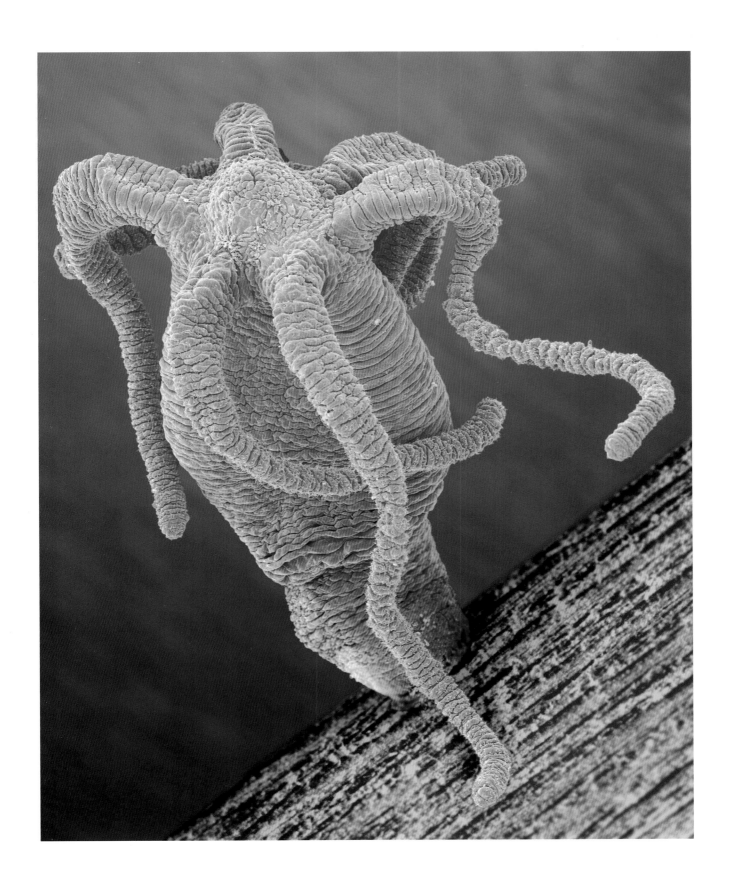

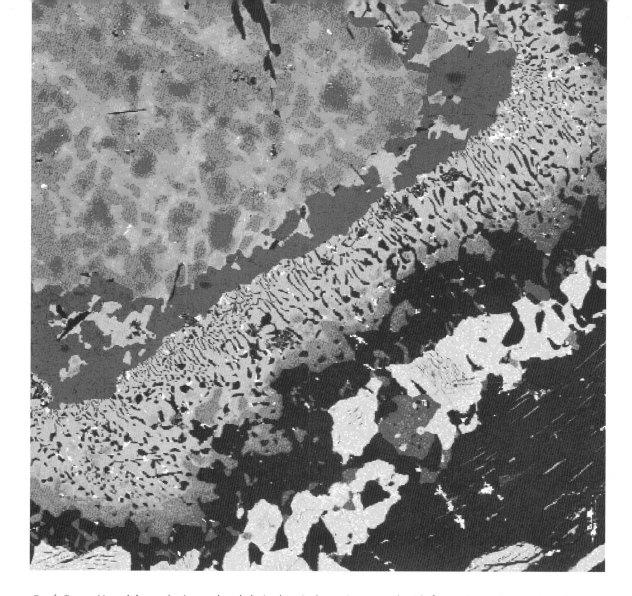

Rock Scan Materials can be imaged and their chemical terrain mapped with fantastic precision using electron microscopy in combination with a microprobe, which can zero in on tiny spots of a sample for chemical analysis. Such a combo stimulates tiny regions of a sample to emit light, which then can be analyzed with a spectrometer, yielding a chemical map of the sample. Here, researchers used such a method to map out the distribution of calcium in a mineral sample obtained north of Stony Rapids, Saskatchewan. Calcium concentrations, which are color-coded, change in a way that indicates seven mineralogically distinct regions in the sample. These range from a relatively pure core of orthopyroxene in the lower right (purple) to a complex of that mineral and two others—magnetite and paglioclase (the middle yellow region stippled with black and red spots) to a region of plagioclase (upper left). Some of the other layers have quartz and garnet. Detailed chemical and physical information about the microstructure of such samples enables geologists to retrace the geophysical occurrences that must have taken place to produce such pastiches.

Using the same technique, a map of magnesium content in a sample from Upper Granite Gorge of the Grand Canyon (right) reveals six distinct mineralogical phases, among them garnet, biotite, and quartz. Among the details that geologists can discern from the data are that this sample never experienced temperatures hotter than 600° C or pressures greater than six thousand times that of atmospheric pressure at sea level. Both images represent sample areas about 2 millimeters on a side (2 x 10^{-3} meter).

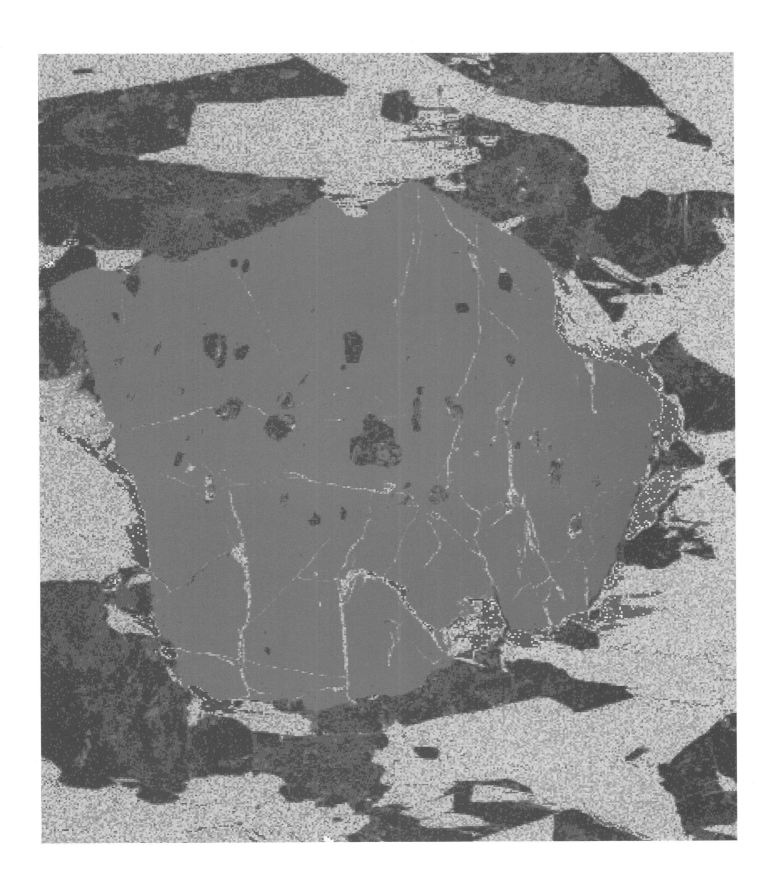

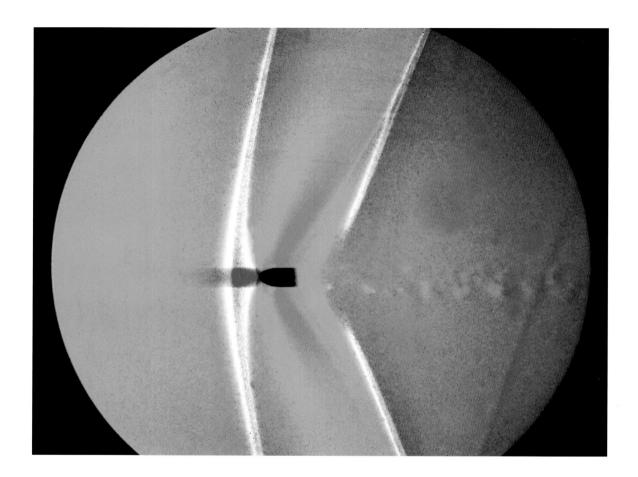

Identified Flying Object Fast-moving objects do a number on the air, but it's very hard to see. One technique for making visible the effects of speed is shadowgraphy. As an object zips through the air, the movement produces various organized and chaotic motions. Light projected through these disturbed regions of air doesn't pass uniformly; the result is a shadowlike pattern. Shown here is a bullet moving faster than the speed of sound, probably over 1,000 miles per hour. The two bands of colors—one leading the bullet and arcing gently to the left and right; the other in two segments behind the bullet, angling more sharply on either side—demark shock waves, or the building up of pressure waves induced by the moving bullet. Directly behind the bullet, what had been a smooth layered flow of air becomes turbulent as it encounters a shock wave. Engineers can use data like this for applications ranging from designing better projectiles to making submarine hulls slide more smoothly through water. The bullet is about 1 centimeter long (10^{-2} meter).

With the help of strobe technology that can produce superfast flashes, ultrafast photography captures details of everyday phenomena like a drop falling into a basin of water. In the example to the right, a crown of water throws tiny droplets upward as the sheet of water in the crown's band shudders and ripples. High-speed photography was pioneered both for technical observations and as an art form in the 1950s by Harold Edgerton of the Massachusetts Institute of Technology. The width of the splash is on the order of 1 centimeter ($\sim 10^{-2}$ meter).

Freezing a Bullet A favorite sport for ultra-fast photography practitioners is to shoot bullets through various things, taking snapshots before, during, and after the bullet does its violence. Here, the photographer captured an image of a bullet just as it severed ten rubber bands stretched directly across its trajectory. As each band was cut, it began retracting, resulting in a series, each member of which is just a little shorter than the next one. The path through the rubber bands stretches about 10 centimeters (10^{-1} meter).

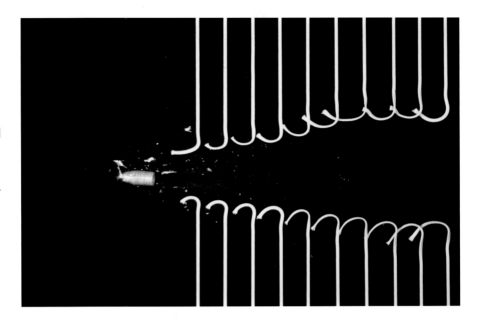

City of Minaturization Nothing is more emblematic of modern technology than the microchip. When integrated circuitry was first invented in the late 1950s, a single transistor took up an area of semiconductor crystal about the size of a fingernail. Ever since, engineers have relentlessly pushed the process known as microlithography and devised new chip architectures and new materials to produce ever-smaller and more compact microcircuitry on roughly the same fingernail-sized piece of silicon real estate. This artificially colored optical microscope image portrays the overall architecture of Intel Corporation's Pentium® 4 chip, one of the most densely packed microcircuitry products available at the beginning of the third millennium. It harbors over 42 million transistors with parts as small as 180 nanometers, not much larger than a virus. Reminiscent of an aerial view of a city, this image shows logic and computation subdivisions, areas for on-chip storage as signals are received and transferred from the chip, and regions with other specific roles. Even with such unprecedented power, the hunger for more music, graphics, and video in computer environments continues to fuel the innovative minds of engineers to develop more powerful chips. A Pentium® 4 chip stretches to about 1.5 centimeters on a side (1.5×10^{-2} meter).

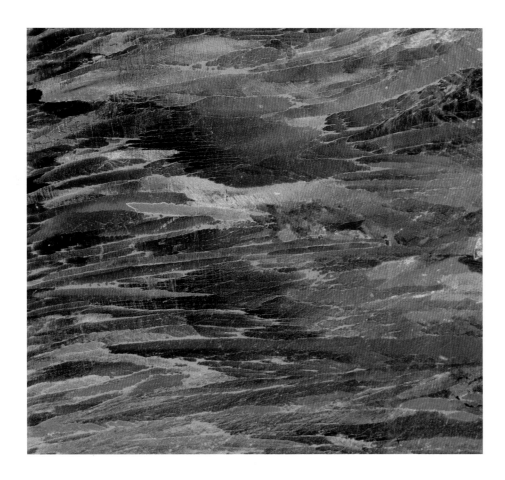

Inside-Out Opal Natural opals do wonders with light. The gems are composed of countless tightly packed mineral spheres, each of which refracts, diffracts, and otherwise manipulates incoming light. The collective optical effect is the iridescence and opalescence that make opals so fitting for jewelry and general wonder. Since the early 1990s, materials scientists—researchers who work to improve the existing materials of the constructed landscape as well as to develop entirely new substances—have known that materials harboring three-dimensional grids of voids, not of solid particles like the opal's mineral spheres, can manipulate light in novel and technologically interesting ways. These materials are called photonic crystals, one kind of which is a so-called inverse opal, pictured here. This one is made of a labyrinthine jungle gym of carbon atoms that form a regular array of minuscule empty cages. As incoming light travels from cage to cage—all of which here are filled with water—it is scattered by billions of walls. The light also is split into intense and clear colors via Bragg diffraction, a process that is basically a submicroscopic version of the one by which a prism separates white light into a rainbow of colors. Photonic crystals have many curious properties, including the ability to allow only certain wavelengths of light to travel through them. They can completely suppress certain ranges, or bands, of wavelengths, opening up such possibilities as making new types of multicolor lasers and building computers that work by manipulating light instead of electrons. The sample shown here is 1.7 centimeters wide (1.7×10^{-2} meter).

Ancient Iridescence Some satellites carry special superfine gratings that split starlight into its component colors. These gratings also diffract light, which is a kind of scattering and reflection that occurs when light interacts with structures such as crystals, whose microstructure is characterized by geometric arrays of rows and/or planes. A result of that added bit of light tweaking is to reinforce certain colors. A similar kind of optical physics takes place in this beautifully photographed remains of an ammonite, an ancient mollusk with a spiral-shaped shell that once teemed in the world's waters. This specimen was thin-walled to begin with, but when it was flattened during the fossilization process, the iridescence of the shell increased. That optical effect is due to close spacing of the shell's layers, which were made of an alternating pattern of the biologically made minerals aragonite and conchiolin. The diameter of the ammonite specimen is about 2.5 centimeters (2.5×10^{-2} meter).

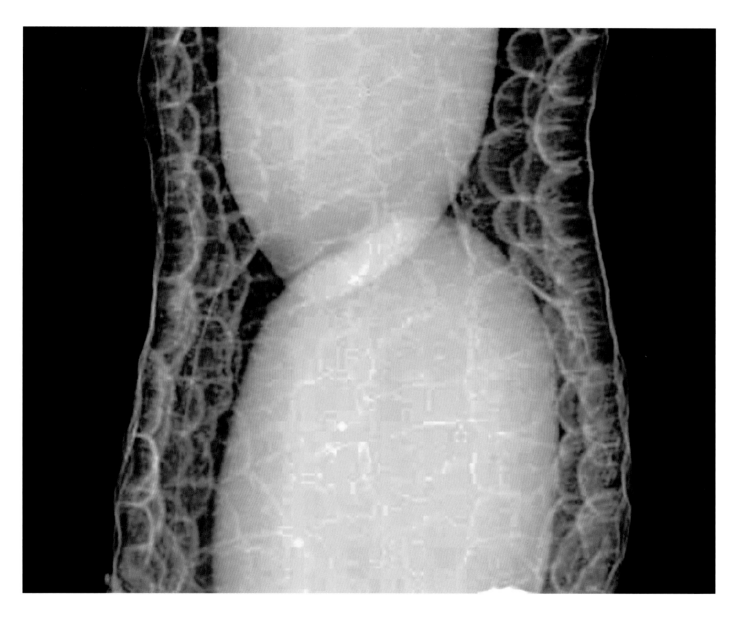

Scanning the Nut CT scans have become part of the general medical experience. "CT" stands for computer-aided tomography; tomography refers to the practice of slicing specimens very thinly, the way a delicatessen slicer slices a salami. In the case of a CT scan, the "slicing" is done by X rays, detectors, and a computer that can parse X rays passing through tissue as though the rays were going through a single slice. The slices can be viewed as a series of two-dimensional images, or a computer can reconstruct those images into a 3-D view. Here, researchers used the CT-scan technique to examine a peanut's interior. The denser nuts inside are surrounded by less dense shell material, which appears almost transparent. Unshelled peanuts typically are 2–4 centimeters long ($2-4 \times 10^{-2}$ meter).

Metallic Frost Watching materials undergo dramatic transformations once conjured thoughts of alchemy. In the absence of modern scientific understanding of the atomic basis of materials, it was quite sensible to think that even lead might be convertible to gold. Watching wood turn into ash and smoke, or ore give way to metal, remains just as magical as ever, but the scientific explanations for what we see open us to an even deeper level of awe. The snowflake-like crystal in this image was created in about ninety seconds by placing a negatively charged electrode into a solution of zinc sulfate sandwiched between two plates of glass. In solution the zinc sulfate separates into its constituents: positively charged zinc ions and negatively charged sulfate ions. The zinc ions are attracted to the negative electrode, a wire at the center of the plates. Once they coat the electrode, small differences in the distribution of electric charge stimulate additional zinc ions to latch onto the growing structure in different places. The process is akin to the way frost forms on a window and the way snowflakes take shape in the atmosphere. The branches of the structure reach out a few centimeters from the central hub ($\sim 10^{-2}$ meter).

Tooth Display Eons ago, in what is now Montana, gomphotheres roamed. These ancestors of modern elephants were usually smaller and had shorter trunks than today's behemoths, but they had longer jaws and tusks. Microscopists at Johns Hopkins University polished the ridged and textured crown of a gomphothere tooth and then photographed it through a microscope. The blue mineral associated with the remaining bits of white enamel is vivianite. The other colors are due to iron oxide minerals that formed throughout the course of fossilization. By carefully studying the ridges and other textures of such artifacts, paleontologists can discern ancient animals' eating habits. Gomphothere teeth could rival the size of an adult human foot. The area shown here is several centimeters across ($\sim 10^{-2}$ meter).

Fluidity To the eye, a clear fluid like water looks homogeneous. Even when it flows and has an irregular surface, it still looks as though there is no real internal structure, but such a structure exists nonetheless. Revealing it takes some cleverness, and the results can be both beautiful and potentially useful. In this computer simulation of a two-dimensional version of a real three-dimensional fluid, faux dye has been added to a fluid undergoing chaotic motion. Rather than immediately spreading throughout the liquid to produce a uniformly colored fluid, the dye first concentrates here and there. Those areas in the liquid where the dye density gradient is the greatest—that is, where the amount of dye changes most drastically in the smallest volumes—are where it becomes distributed into ribbons. In addition to basic investigations into the behavior of fluids, this type of research has applications ranging from industrial process control to optimizing combustion in engines. The simulation is valid at many scales—whether it's one square centimeter like that of a postage stamp or a thousand times bigger—so long as the fluid velocity in the simulation is increased along with the area. Think of the area shown here as comparable to that of your palm ($\sim 5 \times 10^{-2}$ meter).

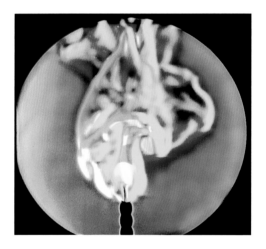

Hidden Flame Finding methods of visualizing flows in transparent gases and other fluids opens otherwise-inaccessible routes to studying the properties and behaviors of these media and their interactions with their surroundings. Schlieren photography achieves this goal by capitalizing on the way regions of a medium that differ in density bend light rays to different degrees. Here, the flame on a small birthday candle generates a lovely and turbulent convective plume, usually invisible. The wild forms in the image derive from a dynamic of rising hot air and the flame's combustion products. The colors, which correspond to directions in which light is bent in the plume, have been added to create a stunning depiction of a candle's hidden beauty. The candle segment and its flame are about five centimeters in length (5×10^{-2} meter).

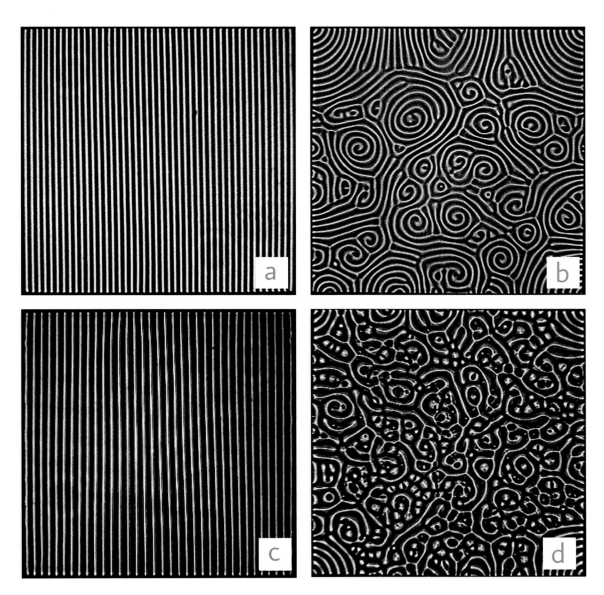

Hunting Patterns From ripples in windblown sand to the stripes of a zebra, patterns are one of the most delightful and intriguing aspects of nature. Many patterns emerge in fluids that are heated or cooled in nonuniform ways, say, because the temperature above the fluid is different than it is below. Some researchers model this situation using compressed carbon dioxide gas that is heated from below and cooled from above. This creates convective motions in which hotter regions of the gas rise (dark) and cooler ones fall (light). Those motions then can be captured with shadowgraphy, which renders visible the changes of temperature in the gas. Shown are four experimental runs. The ones with regular stripes (a and c) are the result of very uniform temperatures on the upper and lower surfaces of the cells. The others (b and d) are just two of the intricate and chaotic patterns that emerge when that regularity is upset to even a very small degree. Chaos researchers strive to uncover simple mathematical descriptions that represent many, or all, possible patterns depending on what the initial conditions of the system are. The width of each image is 6.3 centimeters (6.3 x 10^{-2} meter).

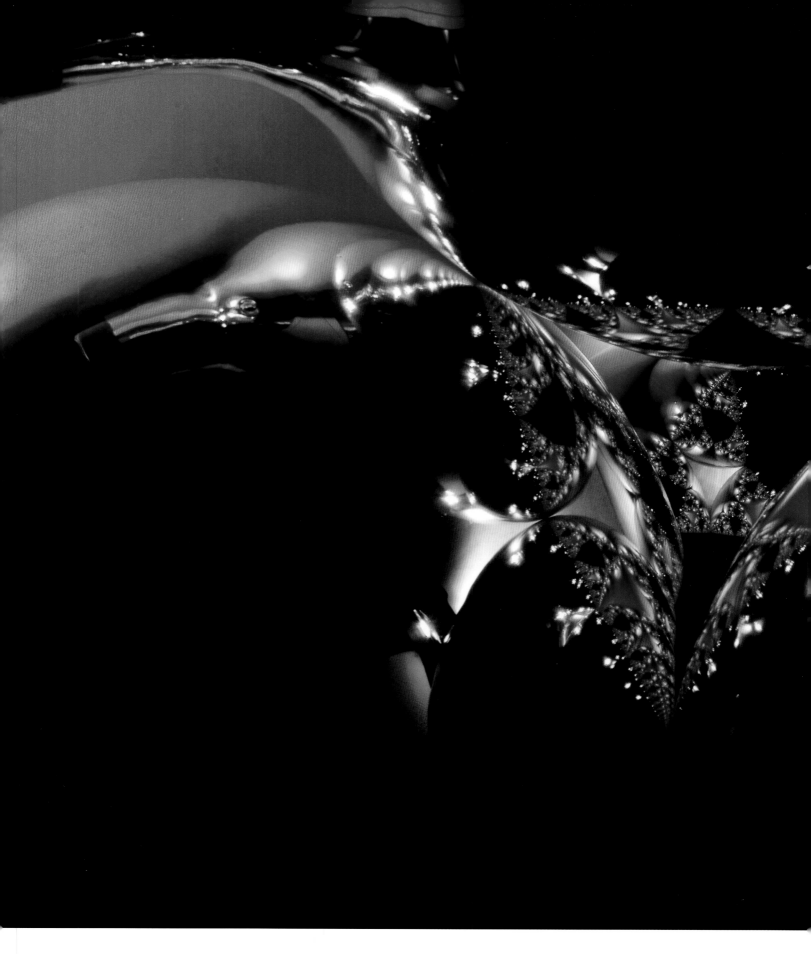

Mathematical Realism

For many years, an interdisciplinary group of investigators has been studying how complex phenomena—ranging from weather patterns to the dynamics of roulette balls—defy specific deterministic predictability yet nonetheless can be precisely defined mathematically. One topic within this field is the emergence of complex forms, such as the shape of a tree or the pattern of grasses in a marshland. Fractal geometry often can be used to describe these shapes and to generate fractal forms in computers. In this image, however, the fractal pattern is created by stacking four mirrored spheres into a tetrahedron (three on the bottom and one on top) and shining a light through one of the gaps into the interior. The colors come from red, white, and blue poster boards that were placed around the structure. Remarkably, this simple experiment replicates some of the most fantastic computer-generated fractal patterns. The structure is a model for what is known as chaotic scattering, which is applicable in dynamic systems such as chemical reactions and celestial mechanics. In the model, the orientation of the spheres determines the way light scatters and reflects and thereby affects the overall pattern. Change the orientation of the spheres even slightly, and the pattern shifts. The spheres are about 8 centimeters in diameter (8×10^{-2} meter).

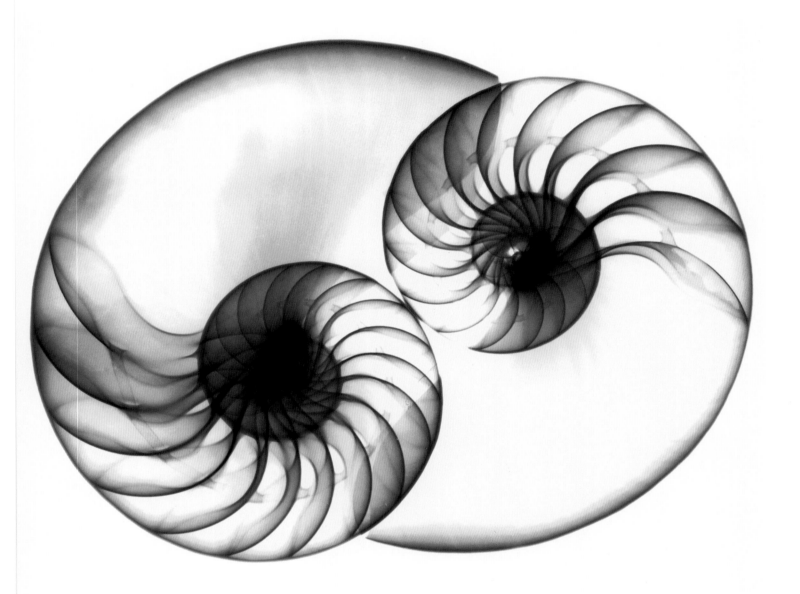

Railing Against Opacity To see usually means that the eyes have harvested light reflecting, refracting, and otherwise emanating from something, whether it is sky, stone, or someone you love. To see with X rays, however, is to see through things, though not entirely. It is seeing through the surface so that the differences in the inner structure and components can become apparent. The most widely known application of X-ray imagery is its use in medical diagnostics. But as artist Albert Koetsier shows with these two images—one of a pair of nautilus shells and one of flowers—the X-ray perspective enables us to see even everyday things with additional dimensions normally unperceived. The shells are about 15 centimeters in diameter, and the flowers are a little longer (1.5×10^{-1} meter).

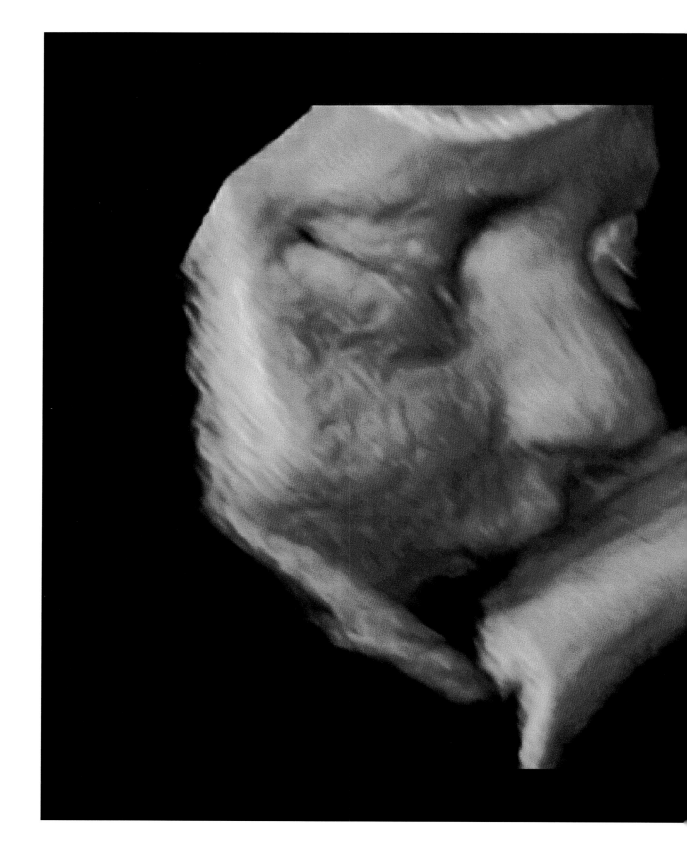

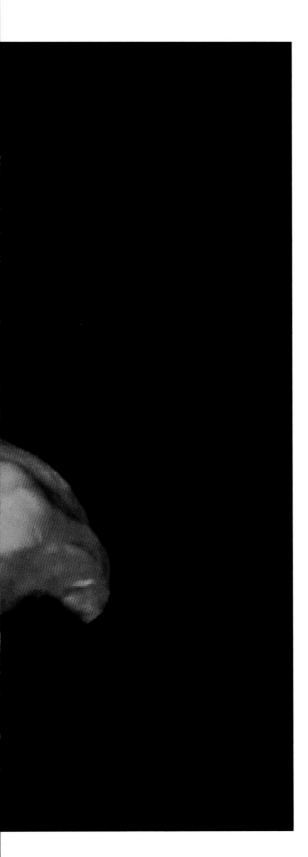

Sound Development For many parents to be, one of the greatest prebirth joys is the sonogram, their first look at the baby. It's the result of a diagnostic ultrasound imaging procedure that shows the developing fetus and helps the physician spot possible abnormalities. A noninvasive procedure, ultrasound uses high-frequency sound to image the body's soft tissue, organs, and blood flow in real time. The images often are ghostly in appearance and require the expertise of a physician or sonographer to discern what is being displayed. Here, with an advanced variant of sonography, a computer synthesizes ultrasound signals into a three-dimensional, sculpture-like portrait of, in this case, a thirty-eight-week old fetal face. The image is about 20 centimeters across (2×10^{-1} meter).

Capturing Lightning Don't try this at home. Take a half-inch-thick block of clear acrylic plastic. Bombard it with electrons moving at the same high speed so that they all penetrate about halfway into the block. Keep it chilled to about −80° C so that the electrons stay put. Now for the shocking part. Take a sharp metal point, even the end of an awl, and hammer it into the plastic. The metal point will serve as a doorway for all of those trapped electrons. In the experiment whose result is pictured here, it took a mere forty billionths of a second for the electrons to drain out, leaving behind the beautiful damage pattern shown. This kind of rapid discharge in a material like acrylic, which doesn't normally conduct electricity, is known as dielectric

breakdown. The process causes local changes in the plastic that collectively show up in a form resembling a bolt of lightning, as it should, since the same principles apply: In an updraft of a thunderhead, positive and negative charges become separated until the voltage difference between the cloud and the ground is so much that the charges leap across the distance between the two. The result, lightning, is natural dielectric breakdown writ large. The kinds of artificially induced electrical patterns depicted here often are referred to as Lichtenberg patterns to honor the early electricity researcher who first generated them in the eighteenth century. The slab of plastic on the left is about 25 centimeters on a side (2.5×10^{-1} meter).

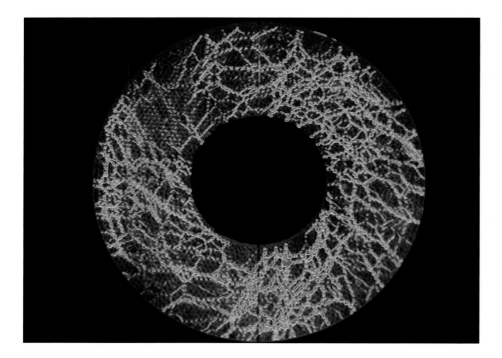

Soft Stuff Many physicists spend their time trying to understand and mathematically codify the complexities that, for example, make materials soft or hard. The way the insides of a material respond to external forces has a lot to do with these properties. Above is an image that depicts the distribution of strain within certain materials. Because the imaging process depends on changes in the way light passes through a material in the presence of strain, the result is known as a photoelastic image, here rendered with false colors that highlight the strain. The "material" in this case is an array of small disks (right), each about the size of a pinky nail, fitted between dinner-plate–sized rings that can twist in opposite directions; it serves as a model system for soft materials that comply when pressed. The set-up enables its users to create shearing forces and to monitor the way those forces redistribute in the disk array. As the disks deform, they rotate the polarization of light passing through them to a degree corresponding to the amount of deformation. These changes are picked up by a detector and then mapped to different colors. The red regions and chains correspond to high local shearing forces and the blue corresponds to regions subjected to lower forces. The data shows, at least on the scale of the experiment, that forces in stressed granular systems—like a pile of sand getting stepped on—are nonuniform and intermittent. The diameter of the outer ring above is about 38 centimeters (3.8×10^{-1} meter).

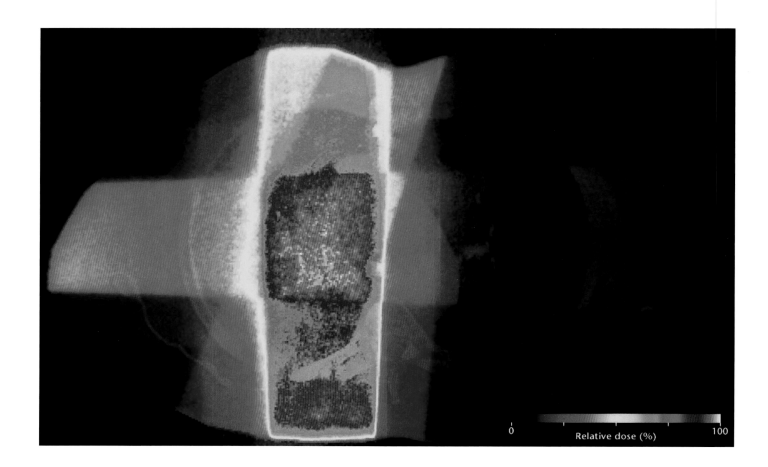

0   Relative dose (%)   100

Damage Control Modern medicine has a powerful armamentarium of tools for treating disease. But the downside of powerful techniques such as radiation treatment for killing cancerous tissue is that their use has to be carefully monitored and managed, lest the treatment itself do more damage than the disease. Toward that end, researchers have developed a computational tool to help oncologists calculate the radiation dose that different parts of a patient will receive as she or he undergoes treatment with several overlapping beams of high-energy photons. The tool is designed to help oncologists assess the relative risks and benefits of the treatment. Shown here is the dose calculation for a treatment involving five intersecting beams converging on a lung cancer. The different colors reflect the different doses received in each part of the body, with the purple marking the highest doses and the red the smallest doses. The body area depicted spans about 60 centimeters (6×10^{-1} meter).

The Look of Tire Noise The occasional car with a broken muffler aside, it's tires that make highways noisy places. As they move, every part of a tire vibrates to differing degrees, radiating noise. Engineers have developed a way to analyze tire designs so they can identify those tire components that generate the most noise. First they measure the intensity of various vibrational frequencies that radiate noise at different points of the tire's treadband, which includes the road-smacking tread and the reinforcing belts. They feed these numbers into a mathematical model, which yields graphs depicting which portions of a tire produce the most intense vibrations as well as an overall fingerprint of the sound the tire produces. Such "fingerprints" of a tire's noisy ways could become useful tools for designers of quieter, next-generation tires. A typical tire's diameter is about 60 centimeters (6×10^{-1} meter).

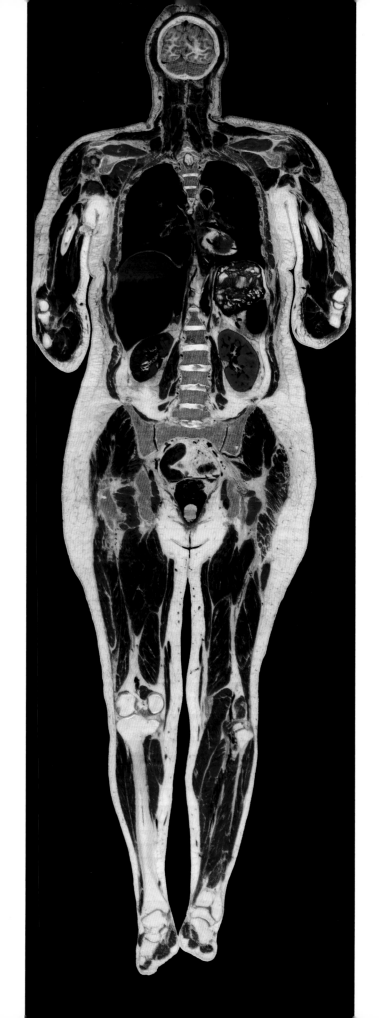

Virtualizing Humanity In 1986, the National Library of Medicine, on the campus of the National Institutes of Health in Bethesda, Maryland, formulated its long-range plan, including "the creation of a complete, anatomically detailed, three-dimensional representation of the normal male and female bodies." It became known as the Visible Human Project. The male body was sectioned into dime-thick sections, which were then photographed and imaged using X rays and magnetic resonance techniques. The woman's body was treated the same way, only it was sectioned at intervals the thickness of file-folder paper. The result is an unprecedented reservoir of data that doctors, biomedical researchers, and their computers can process and manipulate in all kinds of ways. Shown here are so-called coronal sections—which go straight down through the body from head to toe along the shoulder-to-shoulder axis—of both the man and woman. This view is possible only because computers can reconstruct it from the thousands of stored transverse images made from the original slices that cut through the body horizontally. The male body is roughly 2 meters tall.

Using these same troves of digitized anatomic data, biomedical researchers can virtually dissect a human body to selectively reveal any body part, organ, or system. Here, the alimentary canal from mouth to anus, along with the organs that participate in digestion, has been reconstructed from the basic data set, which consists of thousands of penny-thin transverse sections that cut across the standing body the way a belt does. After asking the computer to edit out information regarding bone, muscle, and any other tissue surrounding the digestion system, researchers were able to extract this idealized view of internal anatomy. The length of the digestion system is about 1 meter.

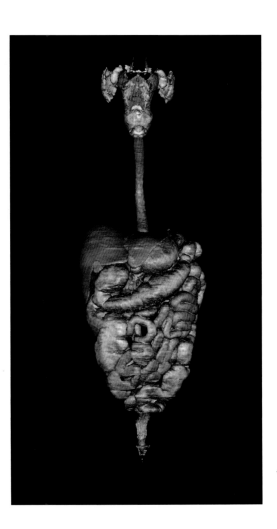

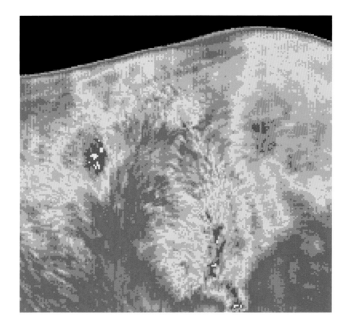

Equine Unease Just beyond the red side of the visible spectrum begins the region of infrared (IR) radiation, which is most often associated with thermal properties, heat. Thermography is a technique in which an IR camera produces an image from the infrared emissions of the scene it is capturing. Here, a veterinarian used an IR system to help diagnose a sick horse. The red/white area on the left corresponds to the location of the kidney and appeared warmer than it should, suggesting that the organ could be infected and inflamed. A subsequent blood test confirmed this hypothesis, and treatment with diuretics, which bring on a so-called kidney flush (a.k.a. lots of urination), cured the animal. In addition to its medical and veterinary applications, thermography is useful in many ways, including monitoring manufacturing processes, detecting electrical or thermal leaks in electronic components, and finding dangerous mechanical strain in structures. The horse flank visible in the image spans about 1.5 meters.

The Cough Finding ways of visualizing flows in transparent gases and other fluids opens otherwise inaccessible routes to studying the properties and behaviors of these media and their interactions with their surroundings. Schlieren photography achieves this goal by capitalizing on the way regions of a medium that differ in their density—because of subtle or dramatic flows—bend light rays to different degrees. One of the most famous Schlieren photographs captures the spreading turbulent flow that unfurls when a person coughs. The person coughing here is professor Gary Settles of Pennsylvania State University, who has used images of this kind in applications ranging from the analysis of nozzles for fire-fighting equipment to the design of better commercial stoves (below). The visible plume of the cougher extends about 1.5 meters; the stove and hood span about 2.5 meters.

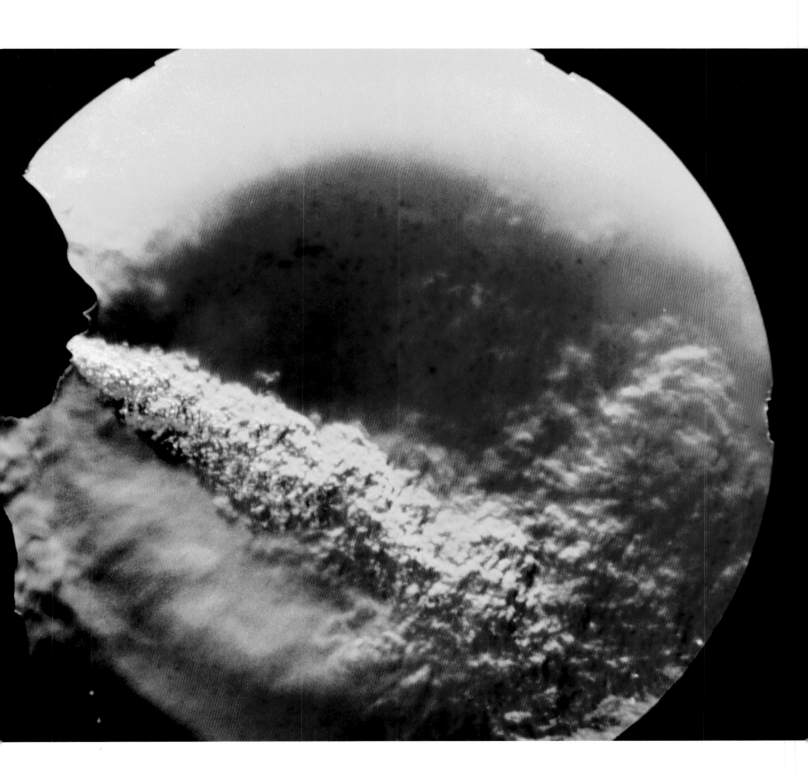

Night Eyes With our color vision we sense sharply: A tiger stands out in the grass, at least in the daytime. When the sun goes down, however, color fades and the world appears black and white. A group of electro-optical engineers at the U.S. Naval Research Laboratory thought it would be great if soldiers could use multiple bands of infrared light to create artificial color vision at night. That way, tanks, enemy soldiers and all other objects might be just as discernible as they are during the day when they're on display in full color. To move toward such a technology, the researchers used an infrared-sensing chip made especially sensitive by chilling it down to frigid temperatures. The chip could distinguish between up to twelve different bands of infrared radiation. By assigning a color to each of these bands, the chip was able to produce artificial nighttime color images of objects, such as this truck, which appeared colorless and ghostly using only a single band of infrared light. In time, the ghostly, green portraiture of nighttime scenes that now characterizes night vision technology could well blossom with color. The vehicle is about 10 meters long.

Sonic Boom Made Visible On July 7, 1999, a Navy F/A-18 Hornet accelerated up to and beyond the speed of sound in the skies over the Pacific. The plane was within view of the aircraft carrier USS *Constellation*, which was carrying Strike Fighter Squadron 151, to which the Hornet was assigned. Aboard the carrier, Ensign John Gay snapped this picture at the moment the aircraft was passing Mach 1, the speed of sound in air—758 miles per hour under normal conditions. It appears that the plane is emerging from nowhere, or perhaps from a hidden dimension of space. The cloudlike formation behind the plane may have been caused by a drop in air pressure, induced by altered movement of sound and other atmospheric waves derived from the plane's motion, causing moisture in the air to condense into droplets. When an aircraft reaches Mach 1, it begins to outpace the sound it produces. The sound waves accumulate in a cone behind the plane and from there develop into a shock wave, which sounds like a boom when it reaches listeners. An F/A-18 Hornet is 16.8 meters long.

Calving Icebergs The name B10 sounds more apropos for an aircraft, but it's the designation given by the National Iceberg Center in 1992 to an enormous quarter-mile-thick iceberg that broke away from an Antarctic glacier several years earlier and then drifted into the Southern Ocean. At about 2,500 square kilometers in area, B10 has been readily visible from orbiting imaging satellites. In 1995, the berg—which rises more than 90 meters above the ocean and floats perhaps as much as 300 meters underneath the surface—split into two pieces. By 1999, the Rhode Island–sized piece designated B10A had edged out of isolated Antarctic waters toward the Drake Passage, which navigators use when sailing around South America's southern tip. As it moved into relatively warmer waters, smaller icebergs have been calving from B10A's edge (upper left) at an increasing rate. These smaller bergs pose threats as they drift into international shipping corridors. As visible from the perch of the Earth-watching satellite *Landsat 7*, this progression of berg generations produces a sparse pointillist portrait of global-scale processes at work. The larger individual bergs are about one kilometer across (10^3 meters).

Farming from the Sky Large-scale applications of water, pesticides, fertilizers, and other agricultural chemicals led to a Green Revolution throughout the last century as farmers' productivity skyrocketed. That success, however, was counterpoised by serious and ongoing trade-offs —the chronic overuse of resources and accumulating environmental damage from the vast amount of agricultural chemicals distributed throughout the landscape by wind and water and diffusion through the ground. One strategy to move away from this agricultural paradigm is known as precision farming, which depends on better information about the condition of specific parcels of land so that harvests and profits can be optimized without overusing chemicals. Overhead imagery is one of the most efficient ways to obtain such information. Here, a NASA aircraft carrying various sensors flew over the Maricopa Agricultural Center in Arizona. The blues and greens in the top image indicate areas lush in vegetation, yellow corresponds to moderate to low amounts of vegetation, and red corresponds to bare soil. In the middle image, which is a map of water deficit, green marks wet soils, while reds and oranges correspond to areas of bare and dry soils. The bottom image reveals those fields that are under the most stress and in need of swift intervention. With such data in hand, farmers can target water, fertilizers, and other treatments specifically to those locations that need them. The plots depicted are about 1,000 meters long (10^3 meters).

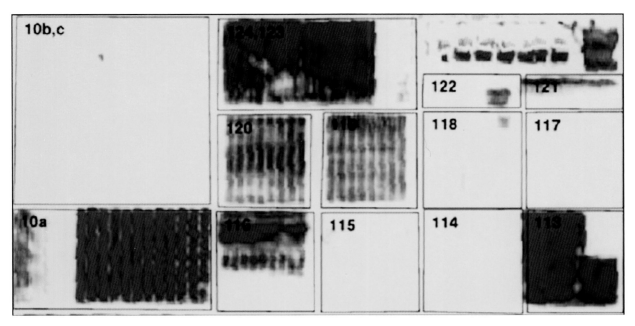

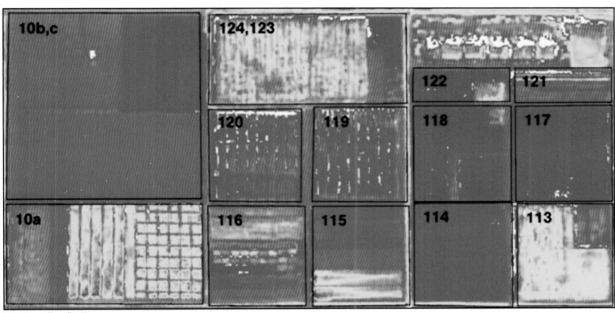

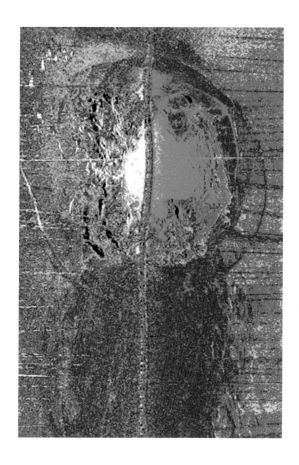

Big Cow Pie Unless you are close enough to use a camera or your naked eyes, sonar is perhaps the only way to look at the bottom of the ocean. By collecting the echoes of sound scattering from the seafloor, the topography and other features of this otherwise invisible realm become visible. Here, acoustics experts beamed high-frequency acoustic signals at a seafloor mud volcano located at a depth of 1,250 meters on the floor of the Norwegian Sea. The signals originated in a "sonar fish" the scientists towed about 40 meters above the bottom. The differing strengths of the echoes returning from different locations were assigned different colors, which reveal different seabed levels and compositions. The amoeboid area in the center of the mud volcano (in blue and green) returned the strongest echoes, an indication of higher density, and the surrounding lower areas (red) sent back weaker ones, which would be expected from softer mud. This volcano is no Mt. Fuji; shaped like a cow pie, according to one of the scientists who has studied the structure, the volcano rises only 10 meters from base to peak. Even so, it represents an important type of geophysical phenomenon that shapes ocean bottoms. The spine-like line running down the middle of the image is a blind spot just underneath the sonar fish. The beardlike formations in purple are probably mudflows that spewed from the volcano no more than a few centuries ago. The diameter of the volcano is about one kilometer (10^3 meters).

Military Omniscience One of the intelligence-gathering systems in the hands of the military is known as JSTARS, for Joint Surveillance and Target Attack Radar System. First deployed in 1991 during Operation Desert Storm, when the system still was under development, JSTARS is a joint Army–Air Force wide-area surveillance system whose primary role is to detect, locate, and track moving and stationary ground targets and to feed that information to military commanders, weapons control systems, and communications and intelligence networks via wire, radio, and computer lines. Central to the system is a cutting-edge, aircraft-carried radar system that can detect and track moving vehicles and even individual soldiers on the ground, much as an airport radar system tracks aircraft. Shown here is a JSTARS-produced image of an unspecified underlying region. Virtually every sizable object that is moving is rendered visible, including those moving off-road. The area depicted here spans a few kilometers (~10^3 meters).

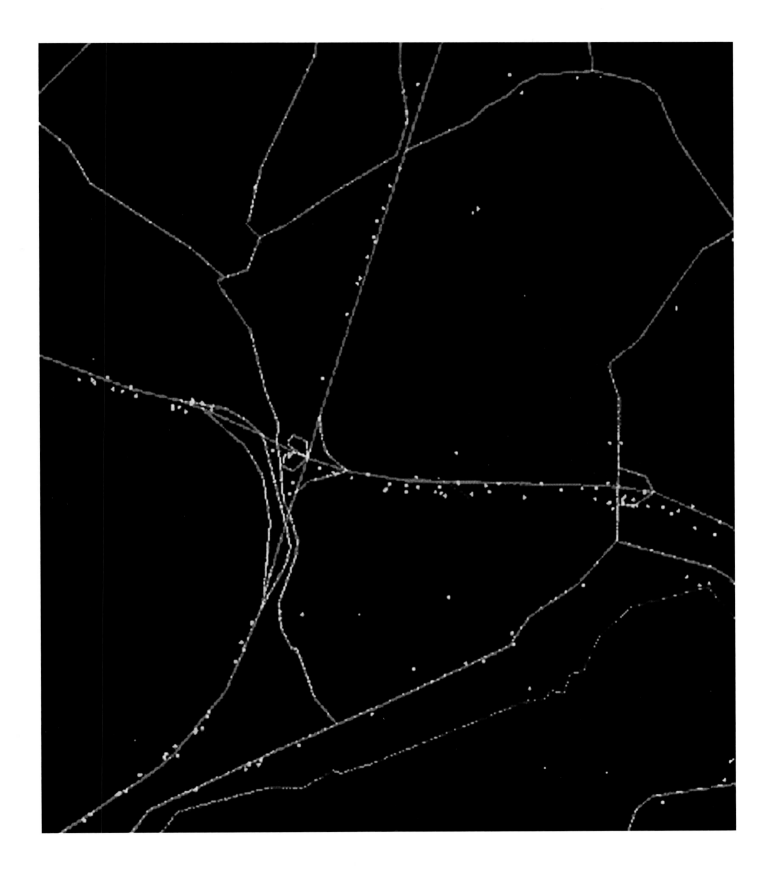

Synthetic Sun Nothing is more exemplary of the awesome power of science and technology than the hydrogen bomb. Energized by the same physics that makes stars shine, nuclear fusion, the vast destructive force of these weapons dominated geopolitical thinking and military strategy for nearly the entire second half of the twentieth century. That strategy, which became aptly known as mutually assured destruction, needs little explanation in the face of such images as this one. This picture was taken on July 2, 1956, at Eniwetok Island in the South Pacific. On that day, the U.S. government conducted a test of a hydrogen bomb whose power amounted to 350 kilotons of TNT. Its classic mushroom shape topped with a roiling fireball remains as terrifying as ever. The test was dubbed Mohawk. The diameter of the fireball spans a kilometer or so ($\sim 10^3$ meters).

Martian Palimpsest Widespread along the Martian equator are terrains that look as though they've been sculpted by sedimentation and erosion processes long observed on Earth. As water levels of large lakes and other sizable bodies of water rise and fall, and as particles in these waters settle to the bottom, layers of various thicknesses form. Over time, these layers stack atop one another, summing into multistepped structures of extraordinary beauty. Late in 2000, researchers working with the Mars Global Surveyor Spacecraft unveiled images of terrain along the Martian equator that share this layered pattern. In this swath in the planet's Valles Marineris, planetary scientists can identify more than 100 individual layers, each of which is about 10 or 11 meters thick. The deeper layers may be as old as 3.5 billion years. These topographies could mean that Mars was once populated by lakes and seas. If that were the case, the argument for the existence of life on Mars— both long ago and perhaps still—would be greatly strengthened. Future missions to Mars may well include visits to these layered locations for searches of ancient fossils. Of course, it is equally possible that some other Martian process, perhaps some kind of seismic settling or wind-driven process, could have produced the fascinating and graceful layers. The long side of the area shown is 2.9 kilometers (2.9×10^3 meters).

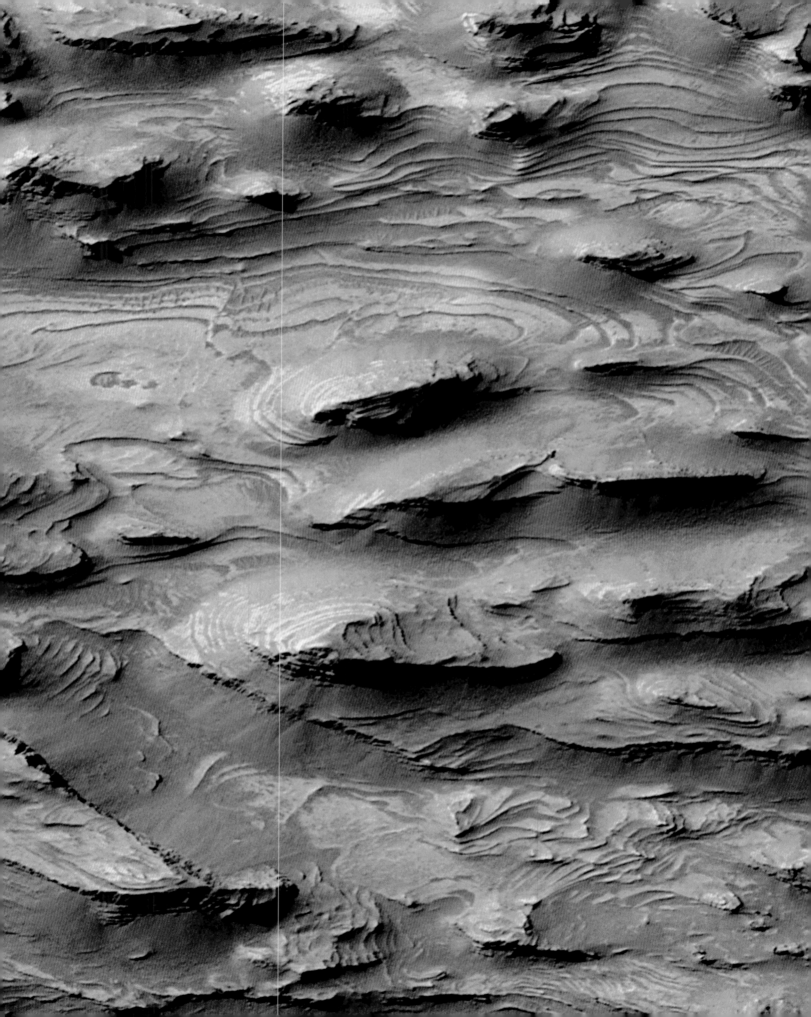

District of Color It takes two simultaneous slightly offset views of the same area to create a sense of depth. In the early 1990s, the Department of Defense's Advanced Research Projects Agency began a program with the goal of developing an aircraft-based system that could generate cost-effective digital terrain information under any conditions of weather, smoke, and lighting. One tool for doing this is known as Interferometric Synthetic Aperture Radar (IFSAR). To create this image of Washington, D.C., in which the colors indicate different elevations, a DC-8 aircraft carried a radar system designed by engineers at NASA's Jet Propulsion Laboratory in Pasadena. To develop the elevation data, the radar's innovative design captured digital images from two spatially separated perspectives. Then by measuring subtle differences in the radar signals associated with corresponding pixels of the two images, the system could infer detailed elevation data. Assigning colors to the different elevations yielded this gorgeous rendition of the nation's capital. IFSAR systems not only can be valuable to quickly develop elevation maps for military planners but can also provide information useful to city planners, road builders, land managers, and others for whom large-area topographic and elevation information is critical. The area shown spans about 10 kilometers (10^4 meters).

Tricorder Wanna-be Human beings are not able to quickly take in large-scale phenomena, such as an entire agricultural region's land use, or evaluate changes that occur over long periods of time, such as the way land use might evolve. To overcome these nonfiction constraints, fictional *Star Trek* scientists carried around book-sized tricorders, instruments that seemed capable of analyzing the chemistry and physics of any object or phenomenon. Early in the 1990s, scientists with the United States Geological Survey developed a program called Tricorder as well as a follow-up program some years later known as Tetracorder, whose mission was to extract valuable geological information from the Airborne Visual and Infra-Red Imaging Spectrometer (AVIRIS). Flown on planes, this instrument can "see" the scene below in 224 colors that span both visible and infrared wavelengths of light. One important application is to rapidly and efficiently assess land use. With the help of reference data acquired from known crops, it was possible to use AVIRIS data to map the vegetation in the San Luis Valley, Colorado. The data shown here were acquired on September 3, 1993. The area spans about 10 kilometers (10^4 meters).

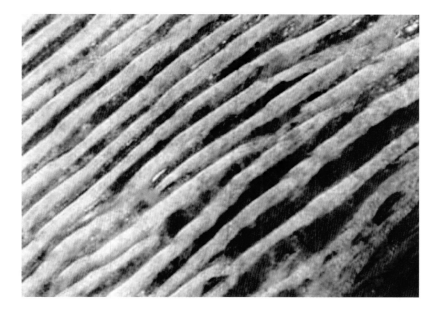

Composition in Dunes The NASA Earth-observing satellite *Terra* carries the Advanced Spaceborne Thermal Emission and Reflection Radiometer, or ASTER. This instrument images the planet mostly through the characteristic ways infrared wavelengths of light are emitted and absorbed by different materials on Earth's surface. By assigning different colors to the different ranges of wavelengths, much as different visible wavelengths appear directly to our eyes as different colors of the rainbow, ASTER simultaneously can capture topographic and chemical information. On June 25, 2000, ASTER imaged a sea of dunes in Saudi Arabia's Rub' al Khali, or Empty Quarter. An iron oxide mineral makes the dunes appear yellow. Clay and silt between the dunes shows up in blue. Located mostly in southern Saudi Arabia but with extensions into the United Arab Emirates, Oman, and Yemen, the Rub' al Khali is the world's largest sand desert, covering an area of 650,000 square kilometers. Its only sometime inhabitants are Bedouin nomads. This small swath of the desert is 28 kilometers long and 37 kilometers wide (3.7×10^4 meters).

Minerals from Above As its name implies, Cuprite, Nevada, is chock-full of copper-bearing minerals (cuprite is the name for a type of copper-bearing ore, and Cu is the symbol for copper in the periodic table), but this area is rich in many other minerals as well. To create this image, using both visible and infrared wavelengths of light, sensors detected the spectral signatures of different minerals. Most of the colors in this image correspond to various iron-bearing minerals, each of which has a specific crystal structure. In effect, images like this one reveal the kinds of chemical bonds that are present in the territory depicted. The absorption data from this site was obtained in 1995 using AVIRIS. Material maps such as this can depict the distribution of minerals, mineral mixtures, vegetation, water, ice, snow, atmospheric gases, and environmental and manmade materials. AVIRIS gathers data in 224 spectral bands in each pixel, which in this case corresponds to a ground area of about 17 by 17 meters. The area depicted is roughly 10 kilometers by 17 kilometers (1.7×10^4 meters).

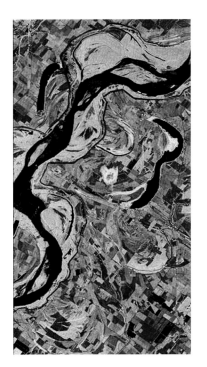

Making Lakes Over the millennia, the Mississippi River has changed course at different places along its length. Some of these course changes led to the creation, through erosion and soil deposition, of what are called oxbow lakes, named after the U-shaped harness worn by the beast of burden. One of these is clearly visible on the right side of the river. This image was taken using the Spaceborne Imaging Radar from the space shuttle *Endeavor* on October 9, 1994, as it flew over the Arkansas-Louisiana-Mississippi state borders. The undeveloped areas along the river show up as pale green while croplands show up in various colors, depending on the conditions of the soil and the type of crops growing. The artificial colors correspond to the different radar channels used by the imaging system. Imagery of this kind can be helpful for discerning flood potential and for helping land managers make more informed decisions about development. The area shown measures about 23 kilometers by 40 kilometers (4×10^4 meters).

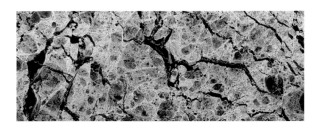

Portrait in Ice In the autumn of 1994, astronauts aboard the space shuttle *Endeavor* pointed a radar camera down at the Weddell Sea in Antarctica. The instrument beamed three different channels of microwave radiation at the ice and then detected and recorded the reflections. Not only are microwaves capable of imaging terrain regardless of cloud cover and sunlight conditions, but they also provide information about its texture, thickness, and other physical properties. With the three microwave channels mapped to the colors red, green, and blue, the resulting images can be simultaneously information-packed and beautiful. Shown here is a 25-kilometer by nearly 18-kilometer swath of Weddell Sea ice. The gray-blue areas mark where there are ice floes more than a half-meter thick, which are relatively solid pieces of ice bounded by thinner ice or water. Within and around these are red-tinged deformed pieces of ice more than two meters thick. Throughout the pastiche, the pattern shifts as floes break apart, exposing new edges of ice to rapid growth due to the frigid air. The dark blue areas are regions of open water, and the light blue lines inside them are due to rough water caused by wind. Any changes in the extent and thickness of polar sea ice harbors clues to the future direction of the globe's climate. This stretch of sea ice spans 45 kilometers (4.5×10^4 meters).

Pastel Volcanoes It took a lot of violent geophysics to produce the Andes Mountains in South America. In this square swath along the Chile-Bolivia border, evidence for this activity is abundant. The image was taken on April 7, 2000, from the orbiting Earth-observing satellite *Terra*, using the Advanced Spaceborne Thermal Emission and Reflection Radiometer, or ASTER. The image combines both visible and infrared data. The dominant feature is the Pampa Luxsar volcano and a complex of lava structures associated with it. On the middle left edge is a pair of so-called stratovolcanoes, which appear in blue due to a lack of vegetation, which appears red in the image. Stratovolcanoes, which include Mt. Fuji and Mt. St. Helens, are steep, conical volcanoes built of alternating layers of lava, ash, cinders, blocks, and other material. The rest of the colors in the image reflect a variety of other surface features, including temperature, elevation, and the ability to reflect or emit light. The area shown here covers a square 60 kilometers on a side (6×10^4 meters).

Face of Europa Jupiter's moon, Europa, represents one of the solar system's best chances for extraterrestrial life because it is covered with a crust of ice. In this swath of crust in the moon's Conamara region, a palette of colors tells a story. The blue and white are due to a dusty layer of fine ice particles that were spewed during the violent event that created the massive crater known as Pwyll, which is roughly 1,000 kilometers to the south. That bombardment of dust, as well as larger debris, left behind small craters and, in places, the crust was even breached. When that happened, water vapor released from underneath the crust carried mineral contaminants aboveground and painted various regions of the surface reddish brown. Undisrupted areas of crust on Europa appear deep blue. The colors in the image have been enhanced to highlight the compositional differences on the surface. The image is a composite made with data acquired by NASA's *Galileo* orbiter spacecraft in September and December 1996 and February of the following year. The area shown is about 30 kilometers by 70 kilometers (7×10^4 meters).

Coastal Soundings Managing a nation's coastal and marine resources is a tough job. Both nature and human beings induce changes in these environments, which can range from beneficial to neutral to catastrophic. It's important to keep an eye on such alternations, since coastal oceans provide means of transportation, commerce, and recreation, as well as food, energy, minerals, and wildlife habitats. The United States Geological Survey (USGS) assumes some of the monitoring responsibilities and the agency uses many tools to carry out that role. The seafloor image shown is part of the Sigsbee Escarpment, which is in the Gulf of Mexico. The image was produced with data collected in the mid-1980s with a sidescan sonar system, which generates images of seafloor topography by bouncing sound off the ocean bottom and analyzing the echoes. These data were combined with bathymetry data, such as seafloor depths, collected some years later, allowing USGS scientists to produce a color-coded map, with the colors representing depths from 1,300 to 3,100 meters. With this sort of data, ocean scientists and geologists can better interpret what is going on at these great depths, and that, in theory, helps the government make better management decisions. The area shown is about 100 kilometers on a side (10^5 meters).

God's Eye View of Fire In 1997, the Pacific island of Borneo was burning, due mostly to agricultural and land-clearing fires that spiraled out of control because of especially dry conditions and an abundance of available fuel from deforestation. The fire spewed enormous plumes of smoke into the atmosphere. On September 22, the National Oceanographic and Atmospheric Administration's Polar Orbiting Environmental Satellite, dubbed NOAA-14 POES, was overhead and snapped this picture. Tracking plumes like this (in yellow) is important for providing short-term health and travel advisories. POES is part of a program to build and use new technologies for gathering long-term data required for environmental monitoring and assessing the effects of global-scale changes. The island, which is mostly obscured by the haze but whose eastern region remains visible on the right, is about 1,000 kilometers wide (10^6 meters).

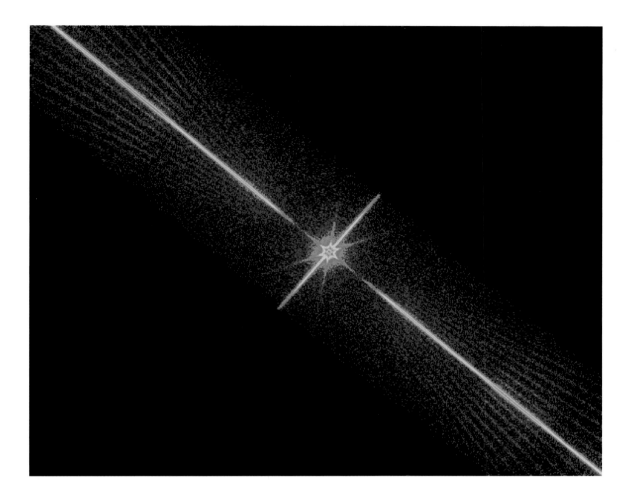

Point of No Return The gravitational pull of a black hole is so forceful that nothing—matter or pure energy, such as light or radio waves—can escape from falling into it. That is why a black hole is totally black beyond its event horizon, the boundary between it and the rest of space. As doomed matter or energy approaches a black hole's yawning grasp, it speeds up and becomes more and more energetic. As it does so, it screams out in the form of X rays. To harvest this image, obtained on April 18, 2000, after nearly eight hours of observation, scientists slipped the equivalent of a prism for X rays just in front of a sensitive X-ray detector at the heart of the orbiting Chandra X-ray Observatory. Known as the Low Energy Transmission Grating (LETG), the gadget spread out X rays arriving at Chandra's mirrors from cosmic object XTE J1118+480. This object is marked by a central black hole with a presumably sunlike star rapidly orbiting around it like a marble in a roulette wheel. The image reveals information about the energy of radiation and about the temperature, composition, and motions of the material falling into the black hole. The highest-energy X rays lie near the center and become less energetic out along the long diagonal line running down from the top left. The bright central hub is due to a small portion of the X rays that are not deflected by the LETG. The bright spokes that meet at the hub, as well as the fan of fainter purple diagonals, are artifacts of the LETG, which is made out of 540 finely ruled gold lines. By combining data from Chandra and other sources, scientists were able to infer that the distance between the black hole's event horizon and the disk of matter falling into it is about 1,000 kilometers ($\sim 10^6$ meters).

Meteorological Composition From above, hurricanes seem only like beautiful dynamic objects, their slowly twirling wisps of clouds obscuring the havoc they may be wreaking underneath. Satellites have been photographing and tracking hurricanes for years and have produced a gallery of stunning images. In this image of Hurricane Andrew in the Gulf of Mexico taken in late August 1992 from the orbiting platform that is part of the U.S. Air Force Defense Meteorological Satellite Program, one of the planet's most awesome displays of power embellishes what appears to be a canvas painted with ocean, land, and sky. Another storm is seen forming in the Atlantic Ocean, at the upper right. Hurricane Andrew caused the deaths of more than 100 people and an estimated $26 billion in damage. The area shown spans about 2,000 kilometers (2×10^6 meters).

Moonview The Sun's corona, which becomes visible to the eye during solar eclipses, is prominent in this image, taken during the Department of Defense's *Clementine* mission in the mid-1990s. Venus, which is closer to the Sun than is Earth, is visible as a small red disk on the right of the image. It looks bigger in this image than it would if someone were viewing it from the vantage point of the *Clementine* spacecraft; the planet is so bright that its light leaked to neighboring pixels of the sensitive imaging chip that captured the image. The glow on the right side of the moon is due to sunlight reflecting from Earth. The moon's diameter is about 3,500 kilometers (3.5×10^6 meters).

Another World's Moon Along its way to an historic rendezvous with Saturn and its moon Titan, the *Cassini* spacecraft sped by Jupiter. The primary objective of the flyby was to get a boost in speed from the huge planet's gravity, but the opportunity to get a good close look at Jupiter was not wasted. On April 20, 2000, a camera on *Cassini* captured this awesome image of Jupiter's ever-fascinating moon Io against the luscious backdrop of chaotic gas that makes up the planet itself. The constant gravitational stresses induced in Io as it orbits its enormous parent make the moon, which is roughly the size of our own, the most volcanically active body in the solar system. Io circles Jupiter every forty-two hours at a distance of about 420,000 kilometers from the planet's center and about 350,000 kilometers from the swirling cloud tops. The heat generated from the moon's volcanic contortions has made this little orb one of the solar system's top candidates for extraterrestrial life. Io's diameter is about 3,600 kilometers (3.6×10^6 meters).

Shoestring Moonshot Completed within a remarkably short twenty-two months from the mission's conception in the early 1990s and expedited at a bargain price of $70 million, the *Clementine* mission to map the Moon in unprecedented scope and detail helped to redefine the "small, better, cheaper" strategy for space science. *Clementine*'s primary goal actually was to test new sensors, imaging technologies, and other advanced spacecraft components for the Ballistic Missile Defense Organization. This image (page 204)—a composite of 1,500 *Clementine* images—is of the Moon's south polar region. Radio-frequency data suggests the presence of ice in the pole's Aitken Basin (the darker region in the center of the image), which is the deepest basin known in the solar system, in some places reaching eight times deeper than the Grand Canyon. The area shown spans about 1,200 kilometers (1.2×10^6 meters) of the Moon's nearly 11,000-kilometer circumference (1.1×10^7 meters).

Venus de Magellan NASA's *Magellan* spacecraft imaged more than 98 percent of Venus during its four-year mission, which ended in 1994. By combining images produced with a radar-based instrument that could detect differences in elevation and some surface chemical composition with additional radar images from the Earth-based Arecibo Observatory in Puerto Rico, this whole-planet image (page 205) was produced. The composite image was processed to bring out small features, and different elevations were mapped to different colors (red is the highest). Some older elevation data for this color mapping came from two other sources, *Venera* spacecraft from the former Soviet Union and those of the *Pioneer* missions to Venus. The diameter of Venus is about 12,100 kilometers (1.21×10^7 meters).

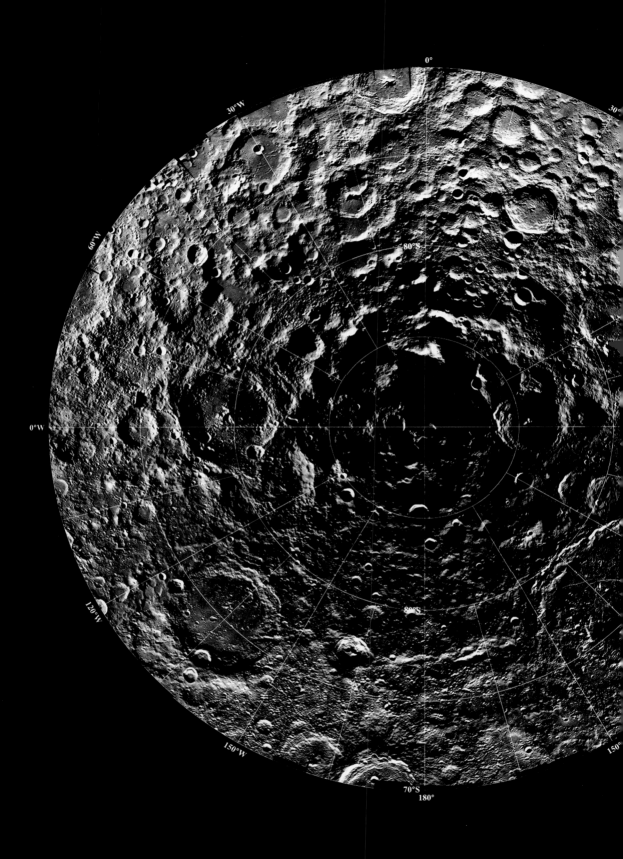

0°

30°W   30°

60°W

80°S

0°W

120°W

80°S

150°W   150°

70°S 180°

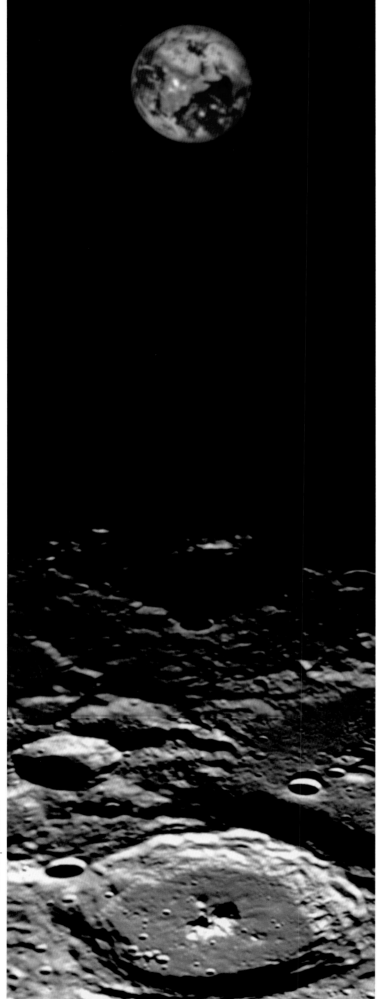

Earthrise One of the most striking developments of the Space Age was the ability to look toward our own planet from the new perspective of elsewhere in the solar system. This colorized image shows the full Earth over the lunar north pole as *Clementine*, a U.S. Department of Defense low-profile military satellite, completed mapping operations during orbit 102 of the Moon on March 13, 1994. It is a clear day over Africa and the Arabian Peninsula. The large lunar crater in the foreground is known as Plaskett. The angular separation between the lunar horizon and Earth was reduced for illustration purposes. Earth's diameter is about 12,600 kilometers (1.26×10^7 meters).

Earthly Plasma In June 2000 the Imager for Magnetosphere to Aurora Global Exploration satellite, known as IMAGE, began fulfilling its mission to gather global-scale data of Earth's magnetosphere, a region of near-Earth space innervated and affected by the planet's magnetic fields. It is here that fields and charged particles from the solar wind interact to create such wonders as "space weather," including the aurora borealis and magnetic storms. While the former inspire awe, the latter have been known to shut down power grids and disrupt satellite-based communications systems. By keeping an eye on the magnetosphere, IMAGE can monitor space weather. As one of the principal IMAGE investigators once put it, "IMAGE does for space what the first weather satellite did for Earth's atmosphere." This particular image of Earth was attained with the satellite's Extreme Ultraviolet Imager instrument. The yellow lobe is due to sunlight scattering from Earth's extended envelope of ionized helium, shown in red, which stretches many thousands of miles into space. Earth's diameter is about 12,600 kilometers (1.26×10^7 meters).

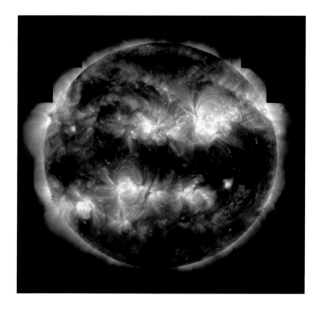

See Spot Spout In the year 165 BC, Chinese astronomers recorded the first observation of spots on the Sun. A follow-up record by a Chinese observer in 28 BC described what must have been a large sunspot as "a black vapor as large as a coin." The fascination with sunspots, which mark areas of the sun with particularly large clusters of strong magnetic fields, has only intensified ever since. In 1998 NASA launched a satellite called TRACE, for Transition Region and Coronal Explorer, whose telescope and CCD detector (charged couple device, which is like electronic film) are trained directly at the surface and near-surface regions of the Sun, the locales where internal magnetic activity shows up in the form of sunspot activity as well as great looping prominences of gas and other dramatic solar behaviors. In the image above, obtained on June 29, 1999, TRACE's instruments captured images of different solar regions in the X-ray wavelength of 171 angstroms (mapped to red), 195 angstroms (mapped to green), and the ultraviolet wavelength of 284 angstroms (mapped to blue). When combined, these three sets of images sum into a stunning portrait of how magnetic fields churn the stuff of the Sun. To the right is a more detailed image of a particularly active region of the Sun. Sunspots range in diameter from 1,000 kilometers to tens of thousands of kilometers ($\sim 10^6$–$\sim 10^7$ meters).

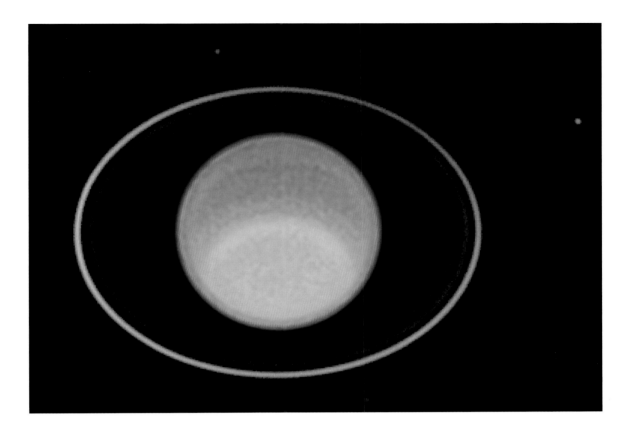

An Eye on Uranus The Hubble Space Telescope, better known for its images of distant galaxies and nebulae, also has been trained on Earth's planetary siblings. With a set of infrared filters, Hubble was able to discern the layered atmosphere of Uranus and some of the planet's inner moons. The infrared image, acquired in 1995, revealed that the atmosphere consists mostly of hydrogen with traces of methane. The red circle marks a thin layer of high-altitude haze. Several of the planet's moons appear as dots beyond the rings. The image is a reminder that Saturn is not the only planet girdled with rings. Uranus' rings, which in reality are as black as charcoal, were rendered visible using image-processing techniques. The diameter of Uranus is about 51,000 kilometers (5.1×10^7 meters).

Solar Tantrums in UV Bright, active regions of the Sun (left) are captured by the Extreme Ultraviolet Imaging Telescope, which is aboard the Solar and Heliospheric Observatory (SOHO)—a joint spacecraft of the European Space Agency and NASA. Shown here are several active regions, home to intense magnetic fields, as sensed in the specific ultraviolet wavelength of 195 angstroms. In such regions, magnetic energy that has built up in the solar atmosphere is suddenly released, launching coronal material outward at millions of kilometers per hour. The Sun's diameter is about 1.4 million kilometers (1.4×10^9 meters).

The Sound of One Sun Vibrating
Every five minutes or so, the Sun
—a dynamic sphere of hot roiling
gas—undergoes an acoustic oscillation, as though it were an enormous bell being struck. Actually,
that's just one of the Sun's 10
million modes of sound wave oscil-
lations, whose wavelengths are
determined by conditions within
the Sun. Over two months in 1996,
an instrument aboard NASA's Solar
and Heliospheric Observatory
(SOHO) gathered the data depicted
here by charting the slight Doppler
shifts in light emitted from the
Sun as it undulates. Called a power
spectrum, it portrays the various
oscillation frequencies in the Sun's
surface as colorful ridges. These
solar motions provide one of the
few means by which scientists
can infer the internal structure
and dynamics of the Sun, whose
diameter is about 1.4 million
kilometers (1.4 x 10^9 meters).

Coronal Glory The Sun's very brightness blinds earthbound viewers to the lovely aura of light that extends outward from its surface. Known as the corona, this veil of light and energy normally only becomes visible to the human eye during solar eclipses, and then never in its fullest extensions. With the help of digital techniques for combining multiple images and for bringing out faint features, this composite of twenty-two photographs reveals the wavy, filamentous corona in stunning detail. The moon normally appears backlit and dark during eclipses, but here digital techniques brought out its faint glow, caused by sunlight that reflects from Earth to the moon and then back to Earth again. The Sun's diameter is about 1.4 million kilometers. The corona extends millions of kilometers from the Sun's surface (~10^9 meters).

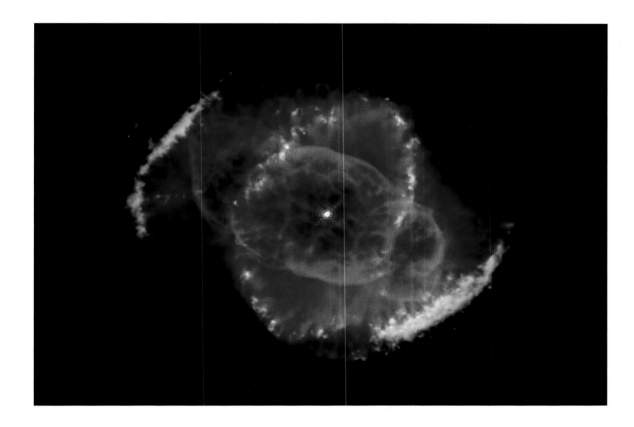

Cosmic Jewels Since the early 1990s, the orbiting, bus-sized Hubble Space Telescope has been supplying the world with breathtaking and scientifically rich images from the near and far reaches of the roughly 14-billion year-old cosmos, which is on the order of 10^{26} meters across. In 1995, one of the most famous images from Hubble was taken. Known clinically as M16, and more poetically as the Eagle Nebula (left), this cosmic neighborhood, which is about 7,000 light-years from Earth, is a birthplace of stars. Within the dark light-years-long columns, dust and cold gaseous hydrogen coalesce and compress enough to become stars that then continue to grow by attracting more and more of the mass around them. These stellar upstarts make them-selves known because they are dense enough for their constituent hydrogen atoms to fuse into larger atoms, mostly helium—a process that yields tremendous amounts of heat and light. A year earlier, Hubble spotted a star representing the other dying end of stellar life cycles. Known as NGC 6543, and more descriptively as the Cat's Eye Nebula (above), the object ranks as one of the most complex planetary nebulae ever seen. Concentric shells of outwardly spewing gas, more concentrated jets of gas, as well as knots of gas produced by shock waves, combined an estimated thousand years ago into this spectacular nebula. At the center, like a tight pupil, is the dying star whose death throes are responsible for the structure. Actually, the various symmetries evident in the nebula suggest that the point in the middle may actually be comprised of two stars orbiting one another. The tallest column shown in this neighborhood of the Eagle Nebula is about 4 light-years tall, or about 4×10^{16} meters. Planetary nebulae, like the Cat's Eye Nebula, have diameters in the vicinity of 1 light-year ($\sim 10^{16}$ meters).

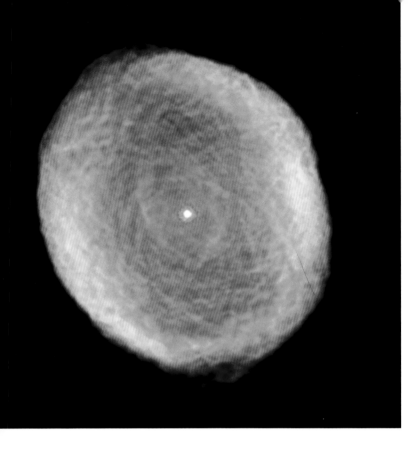

Transient Luster Much closer to Earth, at a distance of about 2,000 light-years, Hubble spied IC 418, otherwise known as the Spirograph Nebula. A few thousand years ago, what had been a red giant star ejected enormous amounts of material from its outer layers. The result was this gorgeous nebula. Remaining from the red giant is a hot core that spews ultraviolet radiation into the expanding sphere of gas, causing the gas to fluoresce much as electricity causes gas to glow different colors in neon lights. This gem of a display is fleeting by astronomical standards. Within a few thousand more years, this nebula will have dispersed and cooled enough that the object will emit far less light. In this image, captured in the year 2000, researchers assigned the color red to light emission from ionized nitrogen gas—the coolest gas in the nebula's outer regions, green to emissions from hydrogen, and blue from oxygen, which is closest to the core and the hottest of the nebular gases. The nebula is about .2 light-years in diameter ($\sim 2 \times 10^{15}$ meters).

Symmetries on Display On January 16, 1996, Hubble captured an image of MyCn18, the technical specifier for a spectacular specimen of a so-called hourglass nebula (right), about 8,000 light-years away. At the waist of the hourglass-shaped remnant of a sunlike star that died slowly long, long ago is what appears to be a stellar remnant that now resembles a gorgeous turquoise eye. In 1997, Hubble spotted and photographed another example of an hourglass nebula, this time with more of a side view (below). Known both as M2-9 and the Twin Jet Nebula, this dying star—about 2,000 light-years away—is not going quietly into that good night: The gas in its twin jets is screaming outward at over 320 kilometers per second, or more than 1,150 kilometers miles per hour. Both nebulae are in the ballpark of 1 light-year in diameter ($\sim 10^{16}$ meter).

A Classic Spiral For good reasons, astronomers have redubbed a galaxy specified as M51 or NGC 5194 the Whirlpool Galaxy (overleaf). In its spiral arms, dust accretes into massive, luminous stars. It's a harrowing and creative time for this galaxy, as nearby galaxy NGC 5195 approaches from a cosmic location just beyond the upper edge of the image. The additional gravitational forces of that galaxy have been triggering the births of stars in the Whirlpool Galaxy, which is about 37 million light-years away. These bright stellar upstarts are visible as clusters of red dots. The disk of the galaxy captured in this Hubble image spans about 65 thousand light-years ($\sim 6 \times 10^{20}$ meters).

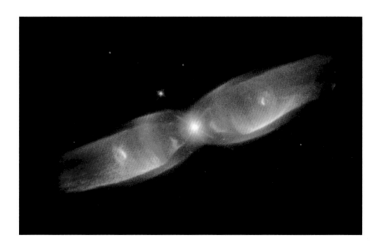

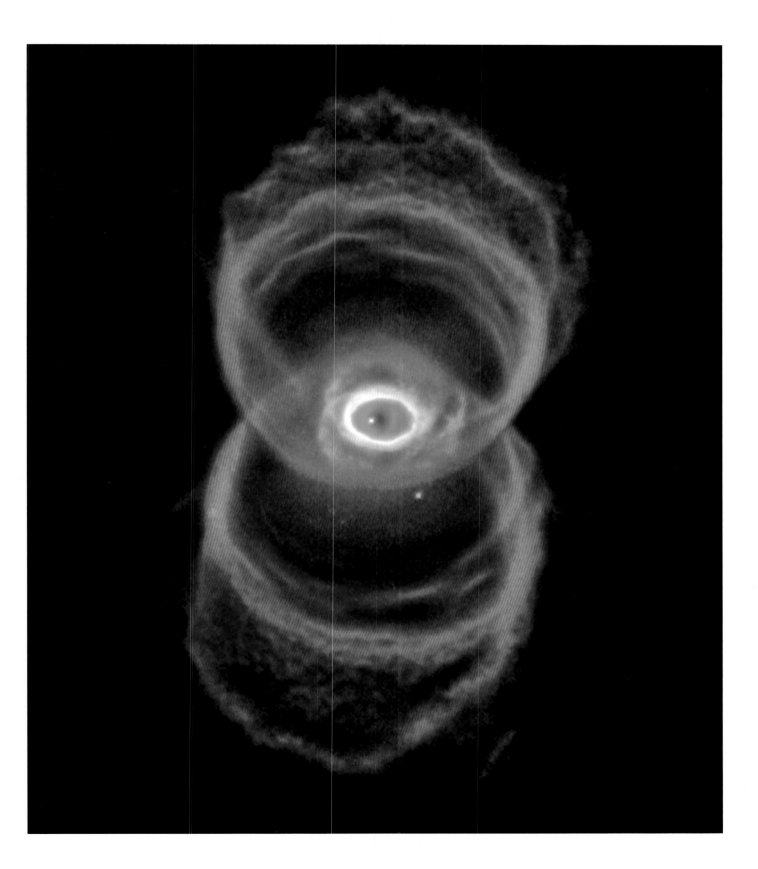

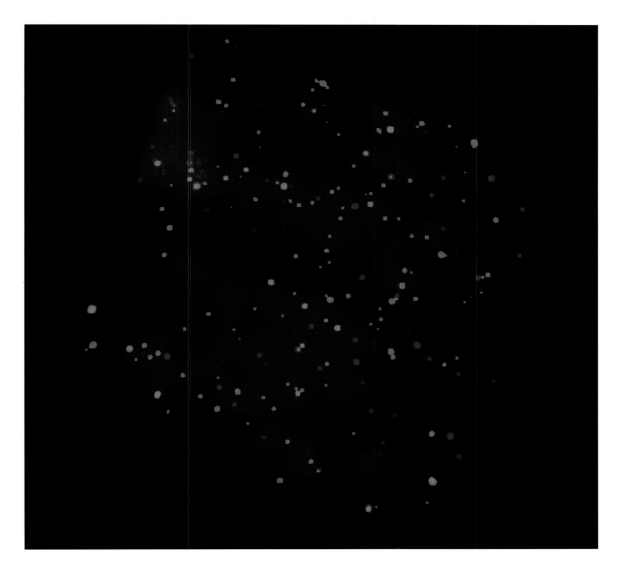

Behind a Cosmic Veil No tool does it all, even multibillion-dollar instruments like the Hubble Space Telescope. A lesser-known orbiting telescope, the Chandra X-ray Observatory, is tuned to see in a way that Hubble cannot. Clouds and dust in the Milky Way block out visible wavelengths, but X rays get through these obstacles and Chandra is designed to capture them. In this speckled image, Chandra was pointed through the so-called "zone of avoidance," which is a cloud- and dust-ridden galactic region impossible to see behind with conventional optical telescopes. Most of the pink and red dots mark X rays from stars within the Milky Way, while the blue dots are due to more energetic X rays beaming in from distant galaxies. Since astronomers were able to identify the distant sources of these dots, they also could infer that the more diffuse energy picked up by Chandra's detectors, and shown here as a dark blue background, derives from gas with temperatures of 10,000,000° C that is concentrated along the Milky Way's plane. It took twenty-seven hours of observation time—over February 25 and 26, 2000—for Chandra to gather the signals that were compiled to form this image. The zone of avoidance is a swath that extends the entire 100,000-light-year length of the galaxy (~10^{21} meters).

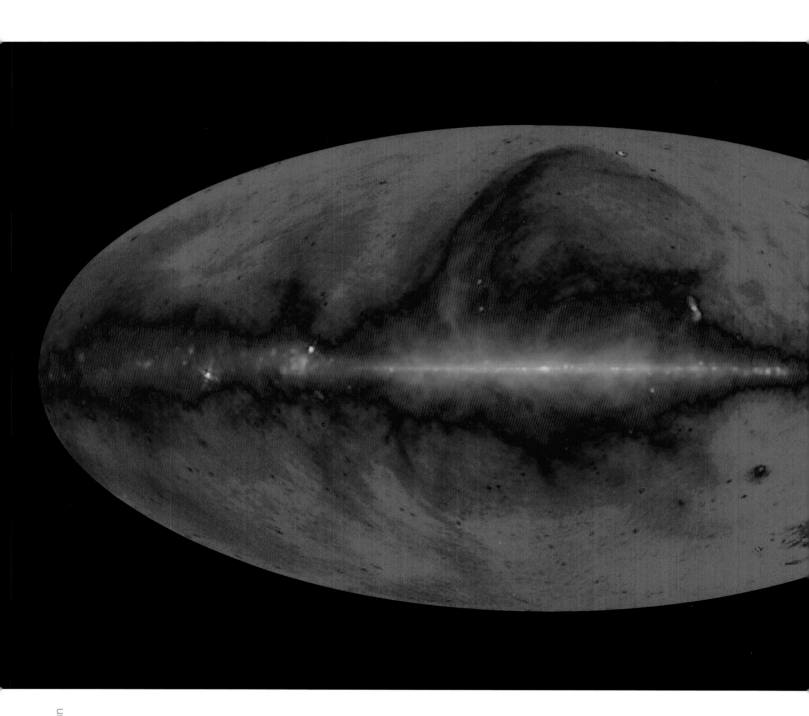

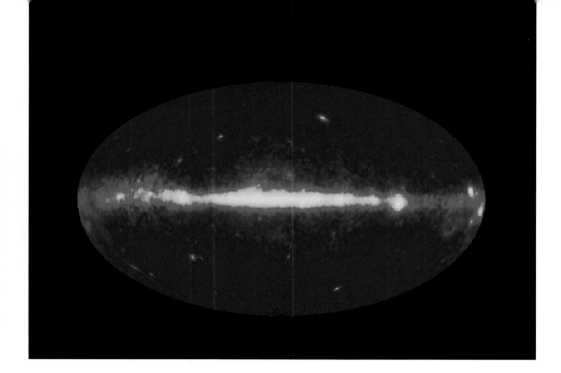

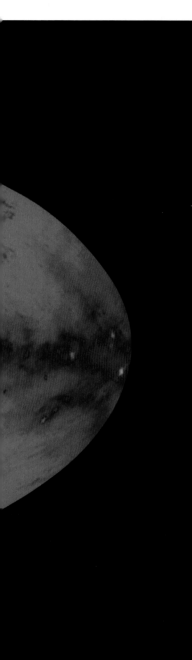

Gamma-Ray Sky It takes violent physics of the most intense sorts to create gamma rays, whose photons carry 40 million times more energy than the photons comprising visible light. The image above comes from NASA's Compton Observatory. As it circled Earth in the early 1990s, a detector monitored gamma rays coming our way from every direction in the Milky Way. The result is an image of the sky as viewed through gamma-ray eyes. The mostly yellow horizontal line corresponds with countless gamma-ray sources in the plane of the galaxy. The brightest spots along the line were made from gamma rays coming from pulsars, which are superdense, magnetized neutron stars spinning at incredible rates. These are made when stars implode with gargantuan force, squeezing most of their matter into a form so dense that a thimbleful of the stuff could weigh a billion tons. The bright gamma-ray spots above and below the central region are due to quasars, which may be powered by supermassive black holes, at the far reaches of the universe. Scientists have yet to figure out the physical mechanisms behind many of the fainter sources of gamma rays. The Milky Way spans about 100,000 light-years (~10^{21} meters).

Universal Radio The night sky is punctuated by starlight that you can see and by starlight that you can't see. In the 1970s, three of the world's largest radio telescopes, enormous curved dishes designed to scoop up radio emissions, listened in on the sky at a frequency of 408 mega-hertz, which lies somewhere between U.S. television broadcast channels 13 and 14. The three antennae had a view of the entire sky—the entire spherical dome as viewed from Earth's surface—so by combining the data and then depicting it in a projection like the one shown, the entire radio sky becomes visible (left). The radio signals in this image (the most intense are colored blue and violet) were due to high-energy electrons spiraling along magnetic field lines, mostly emanating from extreme venues like supernova remnants, pulsars, and star-forming nebulae. The plane of the Milky Way runs horizontally through the center of the image. Emissions from a few other galaxies are visible. For example, the small bright radio spot above the center plane and a bit to the right where the central blue bulge starts descending is due to signals from Centaurus A, one of the galaxies closest to the Milky Way. Our galaxy spans about 100,000 light-years (~10^{21} meters).

Cosmic Sombrero In the late eighteenth century, the French astronomer Charles Messier observed what he described as "a very faint nebula." It was the 104th object in his catalogue of nebulae and so became known by astronomers as Messier 104—until, that is, scientists got a closer look. That's when Messier 104 also became known as the Sombrero Galaxy. On January 30, 2000, astronomers from the Kapteyn Institute in the Netherlands pointed the Very Large Telescope (VLT), a powerful instrument in Paranal, Chile—and part of the European Southern Observatory's operations—at the Sombrero Galaxy. Two years earlier, VLT had been upgraded with the Focal Reducer and Spectrograph (FORS1), a 2.3-metric-ton gadget designed specifically to analyze light from the dimmest and most distant objects in the universe. A composite of three images using FORS1, this image reveals the Sombrero and its dusty rings with a sense of depth and details never discerned before. The central bulge consists mostly of mature stars, and the galaxy's disk, seen here nearly edge-on, is composed of stars, gas, and dust. The galaxy is within the constellation Virgo, about 50 million light-years away, but it's seen here as though the viewer were a mere 170 light-years away. The galaxy has a diameter of roughly 140,000 light-years ($\sim$1.3 x 10^{21} meters).

A Big Picture Using data from the Hubble Space Telescope and other sky-watching tools as input to a model of how the universe has evolved, a group of researchers—known as the Virgo Collaboration—conducted one of the largest computer simulations ever. It took a 512-processor supercomputer about seventy hours to process nearly one trillion bytes of data to create the image of which this is a part. This section depicts a foamy structure with regions of relatively empty space—voids—weaving in and around relatively mass-jammed regions rife with gas, stars, and galaxies, which show up here as a red filamentous structure. The area of this portion of a simulated universe spans about 100 million parsecs ($\sim$3 x 10^{24} meters).

A Most Distant View No image could be more humbling than this famous image from the Hubble Space Telescope. During a ten-day stretch ending on December 28, 1995, the Hubble's Wide Field and Planetary Camera 2 (WFPC2) stared at a tiny keyhole of space covering an area of sky equivalent to about one-thirtieth the diameter of the full moon. This was narrow enough that the few Milky Way stars in the foreground of this little ray of space were far outnumbered by a morass of more than 1,500 distant galaxies representing a huge diversity of sizes, shapes, and ages. Had Hubble focused in any other little keyhole of space, it most likely would have captured a very similar menagerie of galactic animals. This particular image is a compilation of 276 individual exposures obtained with the WFPC2. The range of colors derives from a superimposition of raw images that were taken in blue, red, and infrared light. Bluer objects harbor younger stars and are relatively close compared to the redder and more distant objects, which host cooler and older stars. The Hubble Deep Field includes objects that may have formed within several hundred millions of years of the universe's birth, which means the distance represented in the picture corresponds roughly to the size of the entire universe (~10^{26} meters).

The Oldest View of All In the beginning, there was a Big Bang. So starts the greatest scientific narrative of our time. In 1989 NASA launched the Cosmic Background Explorer (COBE) on a four-year mission to measure and map the ultra-low-energy relic radiation from what had been the most creative explosion imaginable, the one that created this universe. First detected in 1965 by scientists at Bell Laboratories who were looking for microwave sources that could interfere with communications systems, this background radiation is in the microwave region of the electromagnetic spectrum. Theorists had predicted the microwave background radiation due to the Big Bang ought to be uniform, but not completely so, throughout the entire universe. Slight irregularities in its distribution, presumably due to slight irregularities in the Big Bang itself, could account for the overall large-scale structure of the universe, which consists mostly of clusters of galaxies separated by vast voids. The three images shown here are depictions of the background temperature, or radiation, of the entire cosmic sky as seen from Earth. The top image, the dipole, shows a smooth variation between relatively hot and cold areas due to the motion of our solar system relative to the distant matter of the universe. To derive the sky map of the background radiation, the dipole needs to be subtracted from the map of total radiation. Same goes for the data of the middle image, which represents a band of temperature signals from the Milky Way. By subtracting those two sets of foreground temperature data, the map of background radiation emerges (bottom). The dramatic color coding is a bit deceptive because the temperature fluctuations across the entire sky are no greater than one part in 100,000 with respect to the 2.73° K average temperature (that's not even three degrees above absolute zero). The universe is on the order of 10^{26} meters wide.

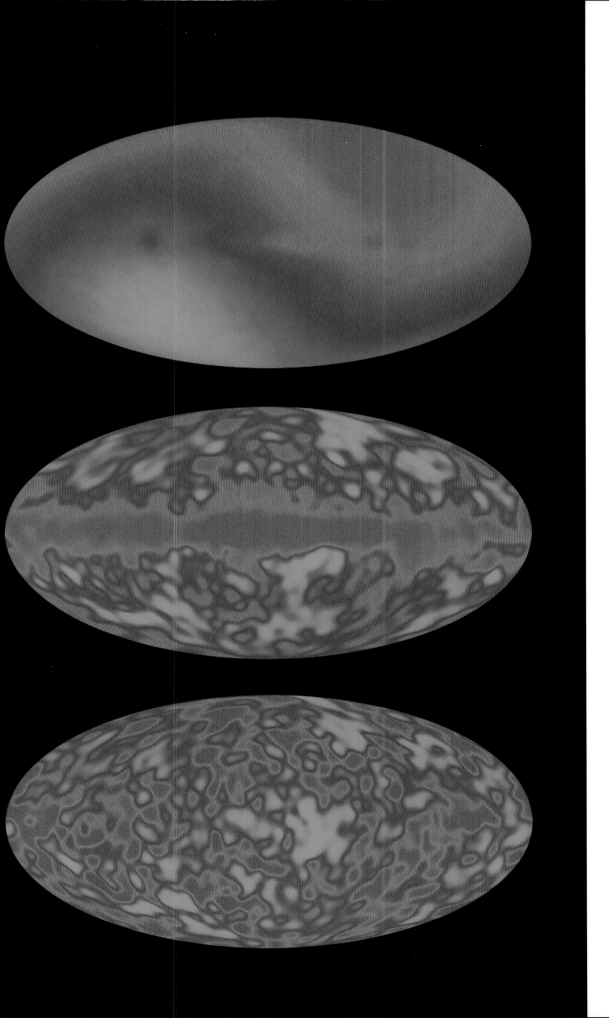

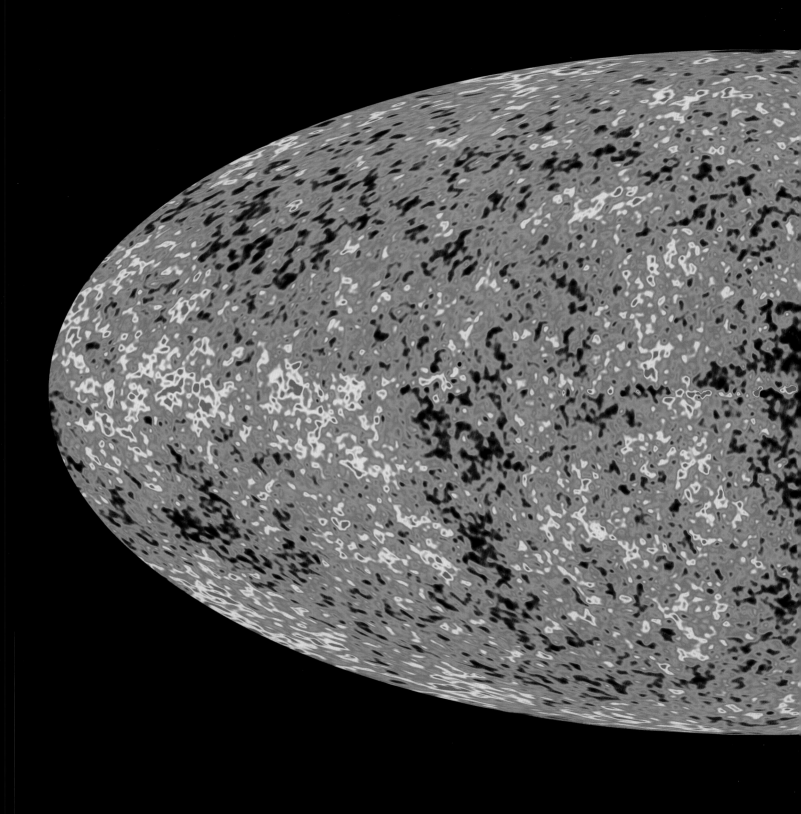

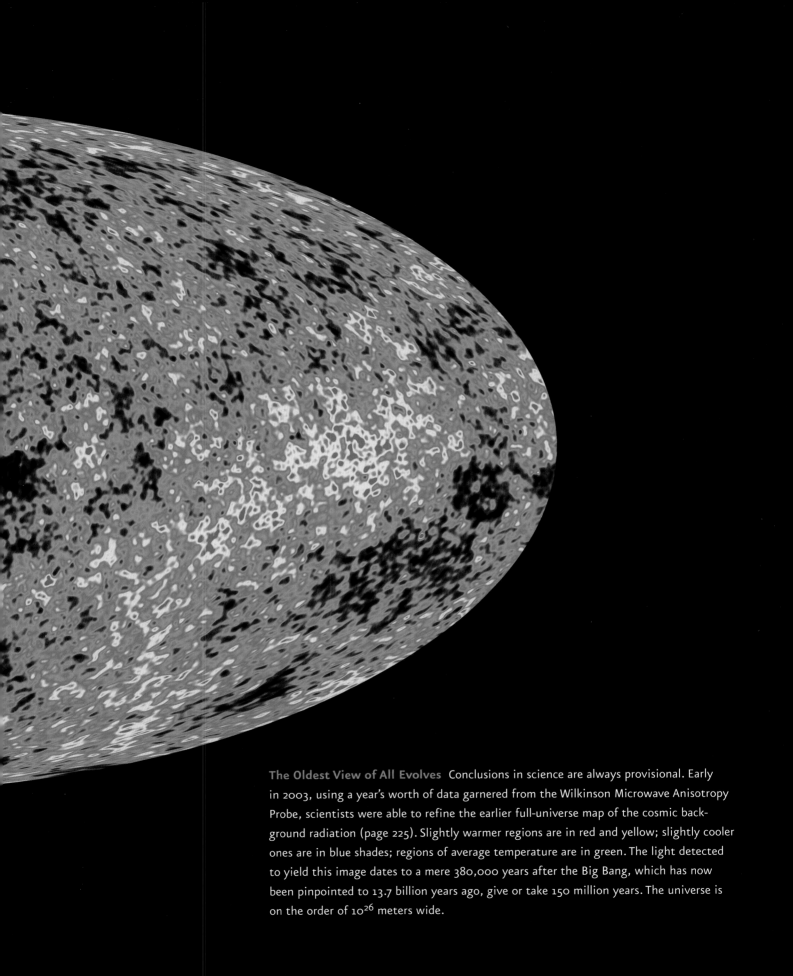

The Oldest View of All Evolves Conclusions in science are always provisional. Early in 2003, using a year's worth of data garnered from the Wilkinson Microwave Anisotropy Probe, scientists were able to refine the earlier full-universe map of the cosmic background radiation (page 225). Slightly warmer regions are in red and yellow; slightly cooler ones are in blue shades; regions of average temperature are in green. The light detected to yield this image dates to a mere 380,000 years after the Big Bang, which has now been pinpointed to 13.7 billion years ago, give or take 150 million years. The universe is on the order of 10^{26} meters wide.

ACKNOWLEDGMENTS

I have had the privilege to be a close observer of the science community for the past twenty years. During that time, I have witnessed the diversification of scientific data and its blossoming into a splendiferous panorama ranging across all categories of phenomena. It is to the scientific community that I owe the greatest debt since it is its tools and data from which the contents of this book derive.

I am grateful to Felice Frankel, whose own seminal efforts to wake the world up to the beauty and power of scientific imagery inspired me in 1999 when I first encountered them, and still does. I am also thankful to David Ehrenstein, who first informed me that Eric Himmel, editor in chief at Harry N. Abrams, might be interested in fielding book ideas from the likes of me. I am especially grateful to Eric for discerning enough substance in my original proposal to accept it and see it through to this handsome volume. For helping me believe I could become a writer, I bow deeply to Holly Stocking.

A cadre of great and generous people devoted a lot of time and effort to the many tasks required for bringing a book like this into the world. Particular thanks go to David Savage, who spent many hours chasing down images and permissions, always with good cheer. In the later stages, Samantha Topol picked up the baton from David with heartening enthusiasm. Designer Helene Silverman took hundreds of written and visual elements and somehow wove them into a beautiful book. Agnieszka Gasparska managed to transform an idea about relating forty-two orders of spatial magnitude into a concrete graphic that I love.

It would take a convoy of trucks to deliver all of the thank yous that I owe to my tireless editor, Sharon AvRutick. Her sensitive and insightful editing, and general oversight of the project, sums into a huge favor both to me and to all of the readers of this book.

I owe a debt also to the scores of people—in various laboratories, public information offices, research centers, and other venues—who tracked down images and made them available to us. There are many hidden from my view who had some hand in making this book possible and so I must pay them tribute blindly.

Finally, I would like to thank my wife, Mary, and my two boys, Simon and Maxwell, whose presence in my life reminds me of what is most important.

I remain solely responsible for any errors of fact or omission that may remain.

INDEX

PHOTOGRAPH CREDITS

2–3: Photograph courtesy of Prof. Mitsugu Matsushita. Contributors to the experiment: Julie Janus, Peter Garik, Trevor Goulette, and Michael Ferretti, Center for Polymer Studies, Boston University/Stan Electrov, Boston Museum of Science. 5: William P. Wergin, Ph.D., Electron Microscopy Laboratory, U.S.D.A. 6–7: Craig Mayhew and Robert Simmon /NASA Goddard Space Flight Center/NOAA National Geophysical Data Center/Defense Mapping Satellite Program Digital Archive. 18: Norman Hathaway. 21: Biomagnetics Working Group, University of Munich. 22: R. Hooke. *Micrographia*, 1665 ©Smithsonian Institute Libraries. 23 top: Anthony van Leewenhoek, Library of Congress, Washington, D.C. 23 bottom: Cajal Institute. 24 top: Ernst Chladni, *Discoveries*, 1787, Musikbibliothek der Stadt, Leipzig. 24 bottom: Deutsches Museum, Munich. 25: Faraday, *Electricity*, 3: pl. III; reprinted in P. M. Harman, *Energy, Force and Matter: The Conceptual Development of Nineteenth Century Physics*, Cambridge University Press, 1982. 28: NASA JPL. 30: CERN Photo. 31:1. Brookhaven National Laboratory. 32: CERN Photo. 33 left and right: Brookhaven National Laboratory. 34: Benjamin Whitaker and Paul Houston. 35: Lisa McDonald/The Institute for Genomic Research. 36–37: Conoscopic images of a quartz platelet taken with the Metripol birefringence imaging system, M. A. Geday and A. M. Glazer, "A new view of conoscopic illumination of optically active crystals," J. Appl. Cryst (2002) Vol. 35, pg. 185–90. 38 and gatefold: Paul Midgely, et al., University of Cambridge. 39: Joseph Michael/Sandia National Laboratories. 40: IBM Research Division. 41: Maria Schnos and Ross Inman, Institute for Molecular Virology, University of Wisconsin. 42: IBM Research Division. 43: John Unguris/National Institute of Standards Technology. 44: W. B. Dress, Jr./Oak Ridge National Laboratory. 45: IBM Research Division. 46–47: G. Medieiros-Ribiero, D. A. A. Ohlberg, and R. Stanley Williams/Hewlett-Packard Labs. 48: Anat Hatzor and Paul Weiss, Pennsylvania State University/Visual rendering by Brent Mantooth. 49: Chad A. Mirkin. 50–51: ©D. Fawcett/Photo Researchers Inc. 52: Jianwei Miao, Pambos Charalambous, Janos Kirz, and David Sayre. 53 and gatefold: Andras Vladar and Michael Postek/National Institute of Standards and Technology. 54: M. Horn-von Hoegen/OMICRON. 55: Dennis Kunkel Microscopy Inc. 56: Prof. Giorgio Gabella, University College of London/Wellcome Trust. 57: Jack Blakely and Kit Umbach, Cornell University. 58: Rev. Mod. Phys.; vol. 71 S288 (1999)/Kwiat and Michael Reck. 59: L. Gurevich, L. Canali, and L. P. Kouwenhoven, Delft University of Technology, The Netherlands. 60: Dennis Kunkel Microscopy Inc. 61: T. Q. Nguyen. 62: Stephen Z. D. Cheng and Christopher Y. Li, University of Akron. 63: Evelin Schröck, Stan du Manoi, Thomas Ried/National Human Genome Research Institute, Bethesda, Maryland. 64: Rafal Dunin-Borkowski, University of Cambridge in collaboration with Prof. Richard Frankel, California Polytechnic State University. 65: Department of Materials Science & Engineering, Royal Institute of Technology, Stockholm, Sweden. 66: Marc Sharp and Kit Pogliano, University of California, San Diego. 67: ©CAMR, B. Donsett/Photo Researchers Inc. 68–69: Dennis Kunkel Microscopy Inc. 70: David Furness, Keele University/Sanger Center/Wellcome Library. 71: Andras Vladar, National Institute of Standards and Technology. 72: ©P. Motta/Photo Researchers Inc. 73: Michael Davidson/Molecular Expressions. 74: Larry Hanke, Materials Evaluation and Engineering, Inc. 75 and gatefold: Eric J. Heller, Harvard University. 76: Peter J. Lee, The Applied Super Conductivity Center, University of Wisconsin. 77: Spomenka Kobe and Zoran Samardzija/Goran Drazic. 78: Karen Larison/Molecular Probes, Inc. 79: Jürgen Berger, Max-Planck Institute for Developmental Biology, Tübingen, Germany. 80 top: OMICRON. 80 bottom: Dustin W. Carr and Harold G. Craighead, Cornell University. 81: Dennis Kunkel Microscopy Inc. 82: Harry T. (Jack) Horner, Iowa State University. 83: Terry McMaster and James Hobbes, University of Bristol. 84–85: James E. Hayden, RPB/Biographics/Nikon's Small World Gallery. 86: G. Vander Rhodes, K. Knopp, M.S. Ünlü, B. B. Goldberg, Boston University Photonics Center. 87: Sanger Center/Wellcome Library. 88–89: Markus Geisen, The Natural History Museum, London. 90: Dennis Kunkel Microscopy Inc. 91: Gerald W. Feigenson, Cornell University. 92: Andrea M. P. Turner, Stephen W. Turner, Harold G. Craighead. School of Applied and Engineering Physics, Cornell University, Ithaca, NY 14853/Natalie Powell and William Shain, NYS DOH, Wadsworth Center, Albany, NY 12237. 93: S. Schmitt, K. Friedrich, Institut für Verbundwerkstoffe GmBH, University of Kaiserslautern, Germany. 94: Jürgen Berger, Max-Planck Institute for Developmental Biology, Tübingen, Germany. 95: Dennis Kunkel and Angel Yanagihara. 96: Dennis Kunkel Microscopy Inc. 97: Molecular Probes Inc./Nataliya Voloshina. 98: Dr. Janet Oliver, University of New Mexico School of Medicine. Access to instrumentation in the Medical School's Electron Microscopy core facility is gratefully acknowledged. 99: All images from Christine Skirius. 100: Richard A. Bley. 102–3: Biophoto Associates/Photo Researchers Inc. 104: Lynn Boatner, S. A. David, and Roxanne Leedy/Oak Ridge National Laboratory. 105: S. Schmitt, K. Friedrich, Institut für Verbundwerkstoffe GmBH, University of Kaiserslautern, Germany. 106: S. Vyas and D. W. Greve, Carnegie Mellon University/Digital Instruments, Veeco Metrology Group. 107: Lucent Technologies' Bell Labs. 108: Image by Dr. Conly L. Rieder, Division of Molecular Medicine, Wadsworth Center, Albany, NY. 109: ©Felice Frankel/J. Aizenberg, A. Black, and G. M. Whitesides. 110–11: Steve Rogers/Imaging Technology Group, University of Illinois, Urbana-Champaign, and Vladimir Gelfand, University of Illinois, Urbana-Champaign. 112: S. Schmitt, K. Friedrich, Institut für Verbundwerkstoffe GmBH, University of Kaiserslautern, Germany. 113: Lynn Boatner/Oak Ridge National Laboratory. 114: Dr. David Furness, Keele University/ Wellcome Trust. 115: Cheema Chomsurin and Scott Robinson, University of Illinois, Urbana-Champaign. 116: Frieda Christie, The Royal Botanical Garden, Edinburgh. 118: Leonard Stern/Lucent Technologies' Bell Labs. 119: Michael Chapman, Murray Barrett, Jacob Sauer, et al., Georgia Tech. 120: Dennis Kunkel Microscopy Inc. 121 left: Robert J. Bick and Brian Poindexter, University of Texas Medical School at Houston. 121 right: Prof. Joseph Perry, Seth Marder, et al., University of Arizona, Tucson. 122: Raija Peura, Institute of Electron Optics, University of Oulu, Finland. 123: David A. Weitz, et al., Harvard University/University of Pennsylvania. 124: Cytomatrix. 125: M. Chaudhury and B-M Zhang/©Felice Frankel. 126–27: James E. Hayden, RPB/Biographics. 128: David Becker and David Rossi, University College London/ Wellcome Trust. 129: James E. Hayden, RPB/Biographics. 130: Raija Puera and Sinikka Komulainen, Institute of Electron Optics, University of Oulu, Finland. 131: Digital Eclipse Gallery. 132: Marna Ericson, Sonny Worel, Maria Hordinsky, Department of Dermatology, University of Minnesota/Nikon 1997. 134–35: Joseph A. Stroscio, Robert J. Celotta, Aaron P. Fein, Eric W. Hudson, and Steven R. Blakenship/National Institute of Standards and Technology. 136: Maria Eisner, Cornell University. 137: D. Walter and C. Meacham, University of Queensland, Australia. 138: Yorgos Nikas/Wellcome Trust. 139: Dmitri Gordienko and Tom Bolton/Wellcome Trust. 140: James E. Hayden, RPB/Biographics. 141: Coral Vincent, Desmond Bradley, et al./*Science*. 142: Ron Sturm/Nikon's 2000 Small World Competition. 143: Eric A. Newman, University of Minnesota. 144: Jürgen Berger, Max-Planck Institute For Developmental Biology, Tübingen, Germany. 145: The Company of Biologists, LTD. 146: Lynn Boatner/Oak Ridge National Laboratory. 147: Charles Williamson, Cornell University. 148–51: William P. Wergin, Ph.D., Electron Microscopy Laboratory, U.S.D.A. 152: Eberhard Bodenschatz, Rolf Ragnarsson, and Brian Utter, LASSP, Cornell University. 153: Dennis Kunkel Microscopy Inc. 154: Christopher Kopf, now at the Department of Geosciences, University of North Carolina. 155: Michael L. Williams, UMASS Geoscience. Images on pages 154 and 155 were acquired on the Cameca SX50 electron microprobe at the University of Massachusetts Department of Geosciences Electron Microprobe/SEM Facility. 156 and gatefold, 157: Andrew Davidhazy. 158: Intel Corporation. 159: A. A. Zakhidov (UTD-Nano Tech) and I. Kharyrullin/*Science*. 160: Norman Barker and Giraud Foster. 161: SkyScan, Aartselaar, Belgium. 162: Photograph courtesy of Prof. Mitsugu Matsushita. Contributors to the experiment: Julie Janus, Peter Garik, Trevor Goulette, and Michael Ferretti, Center for Polymer Studies, Boston University/Stan Electrov, Boston Museum of Science. 163: Norman Barker and Giraud Foster. 164 top: F. Varosi, T. M. Antonsen and E. Ott, University of Maryland. 164 bottom: Gary Settles. 165: Eberhard Bodenschatz, David Egolf, Brendan Plapp, and Reha Cakmur, LASSP Cornell University. 166–67: D. Sweet, E. Ott, J. A. Yorke, D. Lathrop, Institute for Plasma Research, University of Maryland at College Park. 168–69: X-rayography by Albert Koetsier. 170: Philips Electronics. 172–73: Ken Brecher. 174–75: Dan Howell and Robert Behringer, Duke University. 176: Lawrence Livermore National Laboratory. 177: J. Stuart Bolton and Yong-Joe Kim, Ray W. Herrick Laboratories, School of Mechanical Engineering, Purdue University. 178–79: Center for Human Simulation, University of Colorado Health Sciences Center. 180 top: Peter Leando. 180 bottom: ©Gary Settles/Photo Researchers Inc. 181: ©1997 Gary Settles, taken in Gas Dynamics Lab of Penn State University. 182: Dean Scribner, Naval Research Laboratory. 183: Ensign John Gay, USS *Constellation*, US Navy. 184: Landsat 7 Project. 185: Susan Moran, Landsat 7 Science Team and the U.S. Department of Agriculture's Research Service. 186: Warren Wood, Peter Vogt, and Rick Hagen, Naval Research Laboratory. 187: Department of Defense/Federation of American Scientists. 188: Los Alamos National Laboratory. 189 and gatefold: Malin Space Science Systems, MGS, JPL, NASA. 190: NASA JPL. 191: R. N. Clark, T. V. V. King, C. Ager, G. A. Swayze at USGS. 192: R. N. Clark, G. A Swayze. 193: NASA/GSFPC/MITI/ERSDAC/JAROS and the U.S.-Japan Aster Team. 194 top and bottom: NASA JPL. 195: NASA/GSFC/ MITI/ERSDAC/JAROS and the U.S.-Japan Aster Team. 196: PIRL/University of Arizona/NASA JPL. 197: USGS. 198: NOAA POES/Hal Pierce & Fritz Hasler, NASA Goddard Space Flight Center. 199: NASA/CfA/J. McClintock and M. Garcia. 200: Lockheed Martin Missiles and Space/US Air Force Defense Meteorological Satellite Program. 201: Naval Research Laboratory/Lawrence Livermore National Laboratory/BMDO/ NASA. 202: Cassini Imaging team, Cassini Project, NASA. 204: Naval Research Laboratory/Lawrence Livermore National Laboratory/BMDO/NASA. 205: NASA/JPL/MIT/USGS. 206 and gatefold: Naval Research Laboratory/Lawrence Livermore National Laboratory/BMDO/NASA. 207: Southwest Research Institute. 208–9: Stanford-Lockheed Institute for Space Research/NASA/TRACE. 210 and gatefold: SOHO EIT Consortium/ESA/NASA. 211: Erich Karkoschka, University of Arizona Lunar and Planetary Laboratory/NASA. 212: SOHO/MDI Medium-1 Program. 213: Fred Espenak/NASA/GSFC. 214 and gatefold: NASA, STScI/AURA/J. Hester, P. Scowen (Arizona State University). 215: NASA, STScI/AURA/J. P. Harrington (University of Maryland), K. Borkowski (North Carolina State University). 216 top: NASA, STScI/AURA. 216 bottom: NASA, STScI/AURA/B. Balick (University of Washington), V. Icke (Sterrewacht Leiden), G. Mellema (Stockholm University). 217: NASA/STScI/AURA/R. Sahai, J. Trauger (JPL). 218 and gatefold: NASA, STScI/AURA. 219: NASA/GSFC/K. Ebisawa, et al. 220: C. Haslam, et al., Max-Planck Institute for Radio Astronomy/Sky View. 221: Energetic Gamma Ray Experiment Telescope Team/Compton Observatory/NASA. 222: Peter Barthel and Mark Neeser, Kapteyn Institute, Gruningen, The Netherlands/ESO. 223 and gatefold: Ben Moore, University of Zurich/VIRGO Consortium. 224: Robert Williams and the Hubble Deep Space Field Team (STScI/NASA). 225: Cosmic Background Explorer Team/WMAPS Science Team/NASA. 226: NASA/WMAPS Science Team.